Unlimited

Distribution Matters

Edited by Joshua Braun and Ramon Lobato

The Distribution Matters series publishes original scholarship on the social impact of media distribution networks. Drawing widely from the fields of communication history, cultural studies, and media industry studies, books in this series ask questions about how and why distribution matters to civic life and popular culture.

Emily West, *Buy Now: How Amazon Branded Convenience and Normalized Monopoly*

Robin Steedman, *Creative Hustling: Women Making and Distributing Films from Nairobi*

Anne Kaun and Fredrik Stiernstedt, *Prison Media: Incarceration and the Infrastructures of Work and Technology*

Lee McGuigan, *Selling the American People: Advertising, Optimization, and the Origins of Adtech*

Jennifer Holt, *Cloud Policy: A History of Regulating Pipelines, Platforms, and Data*

Aymar Jean Escoffery, *Reparative Media: Cultivating Stories and Platforms to Heal Our Culture*

Rahul Mukherjee, *Unlimited: Aspirational Politics and Mobile Media Distribution*

Unlimited

Aspirational Politics and Mobile Media Distribution

Rahul Mukherjee

The MIT Press
Cambridge, Massachusetts
London, England

The MIT Press
Massachusetts Institute of Technology
77 Massachusetts Avenue
Cambridge, MA 02139
mitpress.mit.edu

The MIT Press would like to thank the anonymous peer reviewers who provided comments on drafts of this book. The generous work of academic experts is essential for establishing the authority and quality of our publications. We acknowledge with gratitude the contributions of these otherwise uncredited readers.

This book was set in Stone Serif and Stone Sans by Westchester Publishing Services. Printed and bound in the United States of America.

Library of Congress Cataloging-in-Publication Data is available.

ISBN: 978-0-262-55271-4

EU Authorised Representative: Easy Access System Europe, Mustamäe tee 50, 10621 Tallinn, Estonia | Email: gpsr.requests@easproject.com

Contents

Acknowledgments vii

Introduction Accessing: Distribution and Mobilities 1

1 Loading: Memory Cards 45

2 Streaming: (Un)bundling and Deep-Edge Computing 83

3 Spreading: Viral WhatsApp and Post-TikTok Futures 117

4 Phishing: Insurgent Trajectories 155

5 Lending: Loan App Scams and Fintech Services 187

Conclusion Limiting: Surveillance, Environment, and Predatory Inclusions 217

Notes 241

Index 295

Acknowledgments

Numerous friends, scholars, and research interviewees have generously offered "unlimited" suggestions throughout the research and writing process. Ramon Lobato read drafts of various chapters and encouraged me to pursue challenging research. Justin Kehoe, Roger Wood, Sudeshna Dasgupta, and Suraiya Jetha patiently guided/edited the book. Marc Steinberg mentored me into platform studies. Julia Velkova, Ranjit Singh, and Lisa Parks engaged me in conversations about distribution infrastructures, users/publics, and media backends. Bhaskar Sarkar, Bishnu Ghosh, Shaina A., Ashok S., Aswin P., Karen Redrobe, Zehra Hashmi, Nikhil Anand, Jenn Wilson, Nicole Starosielski, and Shannon Mattern read book chapters and provided insightful suggestions. During fieldwork site visits, many research participants kindly offered their time and perspective: I am especially grateful to Raghuvendra Prajapati, Seher Bedi, Shadma Shaikh, Deepu S. Edmond, Beni Chugh, Danish Khan, and Prakash Advani. Tejaswini Ganti, Smith Mehta, Seyram Avle, Fathima Nizaruddin, and Tupur Chatterjee generously facilitated conversations among their research networks. For talk invitations to present book chapters, I am indebted to Gayatri Spivak (Columbia), Nona Schulte-Romer (Humboldt), Marwan Kraidy (NUQ), Niels Van Doorn (University of Amsterdam), and Francis Cody (UofT). My deepest gratitude extends to three anonymous reviewers of this book.

Thanks to faculty and graduate students in the Cinema & Media Studies and English departments at the University of Pennsylvania for their collegiality. SAS deans provided research funding. For their care and companionship at a difficult time in my life, thanks to Ian, Juan, Rohan, Ankur, Eszter, Filippo, Pearl, Ari Da, Rubun Di, Josh, Meta, Nicola, Sandeep, Wazhmah, and Sima.

I cowrote two papers with Abhigyan Singh, "microSD-ing Mewati Videos" and "Reconfiguring Mobile Media Assemblages," for the Asian Video

Cultures collection (Duke University Press) and *Asiascape* journal, respectively. Portions of these papers in modified form appear in chapter 1 of this book, and reprint permissions were sought and granted. The same goes for my individually written paper "Mobile Witnessing on WhatsApp" (*South Asian Popular Culture* journal), portions of which are reprinted in one section of chapter 3 (full citations can be found in chapters 1 and 3).

My sister Rohini and my three-year old niece Inaya are a joy and are my key support system along with my father, the most compassionate person I know. This book is dedicated to my mother, Shyamali, who passed away some years back. She is always with me.

Introduction

Accessing: Distribution and Mobilities

In a Vodafone/Hutch phone ad released in India in 2005, the gifted Bollywood actor Irrfan Khan posits a situation where a lover needs to patch things up with his girlfriend on his mobile phone but has no talk time left. Hoping that a simple "I love you" will suffice, he can opt for a 10-rupee *chota recharge* (small recharge).[1] Mobile phone conversations involve intimacies and affectionate expressions, but in 2005, these had to be calculated, budgeted, and rationed. Lovers' exchanges tend toward indulgence and spontaneity, but even such intimate communication could not escape capitalistic accounting. Vodafone ads of the time emphasized economizing talk times.

This discourse was transformed by the arrival of the telecom company Reliance Jio in 2016. Jio promised free talk time and boundless data connectivity at affordable prices. Through 4G (fourth-generation) services, it guaranteed "unlimited" data streaming possibilities. Jio customers did not have to count how many times they were downloading a Bollywood film song or streaming a cricket video. A Jio ad from late 2016 ends on a woman with outstretched hands in an open sun-bathed field as Bollywood film star Amitabh Bachchan's inimitable baritone voice offers the punchline: "Ab Ginh ke nahi, khul ke jio" (Now, do not count pennies, live your life fully). Lovers are apparently now free to chat as long as they want.

The short video platform content creators I interviewed in several Indian towns like Kanpur and Gaya often thanked Jio for bringing *data kranti* ("data revolution") and indirectly helping them have self-enterprising, although precarious, jobs. The above-mentioned examples from 2005 and 2016 might suggest sharp disruptions, but there are continuities in this mobile digital story of India consisting of slow, gradual changes. At one point, there were charges even for incoming calls or a higher charge for calls made and received while roaming in a different Indian state than the person's state of residence. There were intense price wars between Idea Cellular, Airtel, and Hutchison as they tried to outcompete one another and reach India's vast population, and over time, rates started dropping for calls and text messages. Through the 2000s in India, when data rates were nonexistent or too high, and smartphones were rare and precious, missed call was a crucial form of communication. Missed call as a media practice can be traced back to the landline telephone but grew in use across Africa and South Asia with the introduction of cheap feature phone sets. The charges for outgoing calls remained prohibitively expensive for daily-wage earners. People varied the kinds of beeps before they hung up, expecting a call back or conveying a relational message about one lover thinking of another.[2] Between 2010 and 2016, the start-up Zip Dial created hotlines for people to send missed calls, and customers could receive the 2010 FIFA World Cup scores or celebrity tweets along with ads for affordable shampoos and Gillette razors. Zip Dial helped launch offline viral marketing campaigns for its clients through the missed call phenomenon, thereby serving as an important reminder that people sometimes like to work/improvise within limitations.[3]

In this book, the titular "unlimited" functions as a descriptor and as a heuristic. Mobile media companies promise frictionless mobility and tout the boundless capabilities of mobile media through faster download speeds, "seamless" streaming services, and a plethora of apps. This tendency to continuously drum up the breaking of barriers and constraints is not exclusive to mobile phone companies but extends further to the wider discourse around technological growth and unfettered capitalism. This is why unlimited also emerges as a heuristic. Even more significantly, the telecom corporations and app developers' characterization of their offered services as being "unlimited" ironically points to the possibility that they might brush up against "limits," whether those involve resolving storage and bandwidth bottlenecks or the challenge of introducing new apps and technologies on mobile phone networks.

As I learned from my interviews with streaming media distribution professionals, various logistical challenges had to be negotiated to deliver streaming services to mobile phone customers who encountered uneven cell phone connectivity, especially when they were consuming video while in transit to or from work on commuter trains and auto rickshaws. The new video-streaming companies in the India of 2016 had to optimize the encoding process (as part of content delivery) through the bit rate ladder process. In this arrangement, videos are packaged in arrays of segments of different quality and resolution. The source content could be shot in 4K with Dolby sound, but if a user is operating under a bandwidth constraint, then to reach them, the 4K has to be downgraded to HD or the sound channel has to be controlled so that, out of seven channels of audio, only five are made available. This might lead to the loss of surround sound, but the audio-video will not buffer. These logistical constraints suggest the limits of streaming.[4]

"Unlimited data" plans offered by telecoms are not really unlimited, in that after a threshold of data usage is crossed, there is a slowing down of internet speed. This process is called "throttling," where there is a deceleration of connection when customers exceed the so-called fair usage limits. So, after reaching a particular data threshold, customers continue to access data files or stream films but at a reduced speed. Whether it is Verizon Wireless in the United States or Jio and Airtel in India, net speeds are at times decreased for customers with unlimited plans who use a lot of data or relentlessly stream videos and connect to congested cell sites.[5] While the internet can be perceived to be a decentralized and distributed system, it consists of tangible infrastructures like software, routers, internet exchange points, and undersea cables, which can be key corridors and nodes of control as well as act as "critical choke points" that can constrict information flows.[6]

Chuck Tryon notes the disparities between the promises of boundless choice and ubiquitous access from mobile digital technologies and the actual user experiences of delivery models constrained by market fragmentation and geo-blocking.[7] "Unlimited" becomes a way to explore the imaginaries of abundance and mobility that are enabled by the contemporary mobile phone, which—as a medium marked by the rapid convergence of technologies—has purportedly broken all limits, moving from basic cellular mobile telephony to a dynamic and ever-mutating mediascape of complex assemblages. In tracing these possibilities, I do not subscribe to boosterism. As I analyze the various case studies of digital streaming media and loan apps in this book, it becomes amply clear that the rhetoric of boosterism has frequently run into challenges of power and capital and that, in many instances, the reality of socio-technical systems does not match the promises. "Unlimited" as a promissory note of telecom and data service companies is disingenuous not only because it occludes certain kinds of logistical constraints but also because in certain cases states-corporations deliberately impose limitations.

While Jio's unlimited data plans might bring information plenitude for some, several regions of India, including Kashmir and northeastern states such as Manipur and Meghalaya, face media scarcity, as they often encounter internet blackouts. These internet shutdowns are an extreme sovereign measure that the Indian government has normalized to curb what it views as socioreligious and ethnic conflicts in these parts of the country.[8]

Mobilities (Un)Limited

I develop the concept of "mobilities (un)limited" to investigate not only the technological possibilities and limitations of storage and distribution afforded by mobile media systems but also the ways in which sociocultural aspirations about freedom and mobility and shared understandings of gender, religion, and caste dynamics and public-private boundaries are touched and recalibrated by mobile phones. In contributing to the narrative of aspirational India, mobile media are not just helping Indians imagine an optimistic future but are also furthering anxieties about new vulnerabilities. These include temporary and exploitative jobs (even in the formal corporate sector), hate-filled rumors (tinged with casteism and religious fundamentalism), and clashing epistemologies (scientific and historical). This simultaneity of possibilities and limitations has brought me to the structuring concept of mobilities (un)limited.

In what follows, I conceptualize mobilities (un)limited by attending to three interrelated aspects.

First, there are the sociopolitical changes in India that shape and are shaped by the shifting affordances of the mobile phone from 2010 to 2025. The affordances of the mobile phone itself have changed as storage and distribution technologies have been transformed and erstwhile separate forces and flows of mobile computing, cloud computing, big data analytics, and social media networks have converged into mobile media apps. The book explores the relation between such techno-material shifts in mobile phones and the changes in agential capacities, attitudes, and political subjectivities of Indian citizens.

Second, digital infrastructures are emerging to move content, data, and money to reach mobile phone users in India. Studying these distribution assemblages involves tracing socio-technical practices of content delivery networks (CDNs) and analyzing the role of formal and informal human and nonhuman intermediaries, including mobile repair shop owners and credit-scoring algorithms in distribution systems.

Third, with mobilities (un)limited, the focus is on the mutual imbrication of aspirational politics and media distribution, particularly around the questions of who, among India's citizens, has access to platforms (WhatsApp, TikTok) and how they can represent themselves and circulate such representations, and who has access to networks of government delivery services and financial technology (fintech) apps and with what consequences? The Indian government envisions the mobile phone as the site of not only media content but also digital payments. With the circulation of digital media content and money through/to our mobile phones, there is also the circulation of ideologies, practices, and expectations, which can potentially change socioeconomic behaviors and media cultures. I argue that the specific designs of distribution platforms enact/impact particular forms and regimes of inclusion/exclusion and access/surveillance, which are negotiated by everyday phone users in their mobile digital practices.

In this book, readers will find "unlimited" as a promissory note walking on two legs, one operating at the techno-economic level and the other unfolding in the aspirational and experiential level. Both these runs of the "unlimited" promise unbounded horizons but are marked by constraints, whether they are logistical limitations or limits of an individual's cognitive capacities or boundaries imposed by social stratifications. These related dimensions of the "unlimited" shape political subjectivities of Indian citizens and are deployed as governance strategies by state-corporations, especially evident in "Digital India," which is Indian Prime Minister Narendra Modi's flagship initiative aimed to connect the vast Indian population together and set targets of "universal mobile access" and "digital empowerment."[9]

Beyond India's cultural context and mobile phones, "we" are living amid a contemporary moment where the future is presented as a stage where we might be overwhelmed by abundance of some things and scarcity of other things, which will test the limits of "our" capacities and endurance.[10] Amid the dangers of climate change and the hoopla around AI, technocapitalism(s) is remarkably adept at reinventing itself, presenting emerging visions of the future and society, asking us to recalibrate what ethical and bodily limits we are ready to cross. An exhibition at the Experimentarium museum in

Hellerup, Denmark, titled "Future Human: What Is Your Limit?," which I walked into in the summer of 2023, asked that jumping to year 2035, how far were museum attendees ready to go with gene editing, emotional technology, cosmetic procedures, and monitoring feelings? To study "unlimited" and "limits" is to grapple with the "agentive potentiality" of such "futurities."[11]

In projecting an absence of constraints, discourses of unlimited data plans or connectivity gesture to the general ambiguous experiences of "information overabundance," where the shifts in the "scale and scope of content available can destabilize prevailing expectations" of self and how one is supposed to socially relate with others.[12] What is specific to "unlimited," or "limits" is the acknowledgment or anticipation of approaching or having crossed a limit line or threshold level toward those feelings of information overabundance or data saturation.

While considering the affective and experiential dimension of "unlimited," I reflect on some of my interviews with investigative journalists who covered the activities of the Hindu nationalist (supremacist) cow vigilante groups in 2015. These journalists asked their local respondents in the Hindi heartland of North India how they were making sense of the incomprehensible violence of mob lynching perpetrated by cow vigilante groups that tore the social fabric of their town. These acts were provoked by fake WhatsApp forwards on mobile phone screens regarding allegedly Muslim cow smugglers or distorted pictures of them storing cow meat in their refrigerators. Local villagers in the Dadri district of western Uttar Pradesh described such communal flare-ups of mob vigilantism as *hadh par karna* (crossing limits). Amid the different context of Pakistan, Timothy Cooper has described *hadh* as a "moral threshold" that can be identified and policed by a community or society.[13] A close synonym of "hadh" is "seema," and its antonym "aseem" or "aseemit" in Hindi translates literally to "unlimited." The term aseemit is often used to advertise bandwidth services in the Hindi heartland of India.

Content creators and influencers on short video platforms bemoaned the excessively shiny lifestyle choices presented on these platforms that were stoking the desires and aspirations of millions of new mobile phone users who with their present meagre incomes had simply no way of enjoying the presented lifestyle. This was another kind of breach of limits, which was described by one self-motivation influencer from Agra, Simran Jain, as the dominance of "way too cringe" content: "Kuch zyada hi cringe hoti ja rahi hain cheezein, har teesra banda yeh dikha raha hain main toh Mercedes le raha hun, arre bhai main toh BMW le raha hain hoon, arre bhai main toh ghar le raha hoon" (Things are becoming way too cringe nowadays. Every third person is flashing Mercedes and BMW, and showing off a newly bought house).[14]

For Simran, whom I interviewed in July 2024, the preponderance of such videos circulating on platforms such as Instagram and Moj created pressure on audiences, given the disparity between incomes and aspired lifestyles. They felt frustrated that they would never catch up with the glitzy world: "Kyunki unke upar bahut pressure aa gaya hain . . . yeh dekh ke ki duniya kitni zyada bright hain, is duniya ke saath hum kaise chal payenge 9-5 job kar ke" (They are facing tremendous pressure by seeing that the world is so bright . . . how will we keep up with this world with our 9 to 5 job).[15] This example points to the irresolvable tension between constraints and what is perceived as freedom, which is at the heart of mobilities (un)limited. Mobilities (un)limited is not trying to balance out, or provide a dialectical synthesis of, cell phone–induced aspirational goals and India's class asymmetries, but emphasizes the irreconcilable contradiction of these two. To invoke Lauren Berlant's "cruel optimism" thesis, despite facing adversities, aspirants keep holding on to the hope of bright futures that seem even more unattainable with each passing day. Mobile phones or mobile circulations are perceived to be both creators of—and bridges across—the gaps between hopeful desires and realized goals.[16]

The book examines a range of case studies that demonstrate not only the infrastructural limits of mobile distribution systems and regulatory limits of governmental policies but also cultural limits induced by social hierarchies. State-based regulators try to catch up with rapid technological changes in mobile media, and in some cases, hackers and scammers find alternative ways of using mobile phone technologies to exceed/trespass beyond the limitations set forth by earlier regulatory protocols. Caste and gender stratifications in India come with social/familial proscriptions on the uses of mobile phones. In this book, terms cognate to "limits" (thresholds, constraints, choke points, edges, and boundaries) and to "unlimited" (free, proliferating, and unbounded) become important. Thresholds/limits can be thought of as drawing a line to create a border or boundary beyond which something/someone may not, or does not, go. At the same time, threshold/limit could also be the line or level that once crossed/trespassed could lead to change.[17] With shifts in mobile phone technologies and mobile media distribution systems, this book deals with examples where phone users or distributors are approaching limits/thresholds, feeling the sensation of being on the verge of profound transformations. These can be a logistical breakthrough or a shift in social-cultural status or reorientation of attitudes. As *Dalit Bahujan* ("lower-caste") women content creators, whose entry into public spaces such as temples and schools has been restricted in the past by social proscriptions, are making their expressive content spread/move on mobile media distribution networks (TikTok/Insta Reels), erstwhile cultural limits are being displaced and renegotiated.[18] The limit line/threshold level at a particular moment might be fixed, but over a period of time, it does not remain set in stone and is itself mutating and processual.

In mathematical calculus, it is often advised that when we see "limit," we should think "approach." Approaching the limit, transposed from mathematical to social calculus, can entail time and work spent in the present as one strives for a desirable future, what some have called "aspirational labor."[19] Approaching the limit involves dwelling in the affective state of anticipation—"an excited forward looking subjective condition" marked by anxieties and yearnings as the individual adjusts to speculative scenarios, hedged bets, and moving aspirational targets.[20] Karishma Desai argues that "anticipation" is a key lens and "affective state" to understand "educational imperatives" (and directives) to aspire in India.[21]

With "mobilities (un)limited" I examine the entanglements of aspiration and anticipation as individuals conditioned by smartphone apps and their distribution logics approach limits.

The Indian government has struggled to generate jobs for the rapidly growing youth population. Unemployed Indian youths sitting in their homes are instructed/encouraged by the state to not be despondent or remain idle, but continue aspiring to a bright future and use their smartphones to enroll in online courses through the Skill India Digital Platform. These state-certified online training courses are a manifestation of the confluence of "Skill India" and "Digital India" initiatives. I found courses related to mobile repairing, WhatsApp marketing, and reels masterclass at the Skill India Digital Hub, which often cater to the digital market and are seen by the government as increasing the chances of youths to find employment or generate self-employment entrepreneurship opportunities.[22] Aspirations of youths, their anticipation about the future amid precarious employment conditions, are shaped by education and technology, and here we see the mobile phone and the government's skill training app as education distribution technologies spreading aspirational pedagogy. Training institutes set up as part of Skill India, writes Karishma Desai, not only provide technical skills training but impart life skills that seek to cultivate affective dispositions like self-confidence and resilience so that trainees continue to aspire amid (despite) the constrained present.[23]

With a focus on the hyperconvergent medium of the mobile phone with its myriad context-specific socio-technical potentials and limitations, this book offers terms like "mobilities (un)limited" and "mobile media distribution" as a way to study the assemblages, (im)mobilities, imaginaries, and distribution supply chains accompanying mobile digital practices. Such practices involve the mobile internet but extend well beyond it. While conceptualizing the mobile phone as part of, and embedded in, Indian cultures, this book investigates how lived experiences of social space and time in India have changed with mobile technologies.

The chapters in the book deal with mobile media practices and phenomena such as up-/downloading, streaming, spreading, phishing, and lending that are sustained by distribution assemblages and networks. Across the various studies of distribution ecologies assembling around mobile media use in India, my effort is not to focus just on technical affordances or just on regulatory frameworks or just on sociopolitical mobilizations but to connect them together. The reason for invoking mobilities (un)limited is to not merely point out that telecom companies are marked by paradoxes of first promising limitless connectivity and then encountering logistical limitations. This book goes beyond such contradictions to draw interscalar and multimodal relations across infrastructural workings and breakdowns and affective registers of expression and experience. Anticipation is not just the affective state of individual aspirants but can be seen as "embedded in socio-technical systems" of distribution infrastructures.[24] Anticipatory logics of distribution infrastructures can be found in the predictive analytics of lending platforms premised on the use of "nontraditional" social media data to prognosticate borrower's financial behavior. Anticipatory modes can be located in the speculative human and algorithmic decisions that are part of content delivery network operations that select particular streaming TV show episodes to put on edge cache servers closer to the user so that such anticipated popular on-demand content does not have to be fetched from far-away cloud data centers during moments of dense data traffic congestion.

Drawing on scholarship in mobile media studies, South Asian studies, media industries, science and technology studies, and media anthropology, I study the process of mobile media distribution in India by attending to the technical specificity of platforms and infrastructures, the political economy of their operations, and the views and experiences of distributors and users. This is based on a multimethod research approach that includes years of engaged fieldwork and site visits to rural districts of Haryana (Mewat), Bihar (Gaya), and Jharkhand (Jamtara) and urban metropoles (Mumbai, Delhi, and Bengaluru); qualitative open-ended interviews with phone users, content creators, talent agents, data scientists, industry workers at different sites of the distribution supply chain; document analysis of regulatory reports and circulars; and extensive reading of financial and technology newspapers, magazines, and industry periodicals. To map the circulation of contents, communications, and payments through mobile media devices, I adopt a multiscalar methodology that is at once concerned with the cultural politics of individual mobile phone use (involving liking, sharing, forwarding, swiping, and scrolling) and the state-corporate regulatory protocols that encourage and constrain such cross-platform dataflows.

While the focus of media studies scholarship historically has disproportionately been on production and consumption, several recent studies have addressed the lacunae of attention to media distribution. Alissa Perren has stressed the importance of a cross-disciplinary dialogue on distribution that reassesses the "space in between production and consumption."[25] Innovative approaches to studying disruptions in news distribution or retail supply chains with increasing digitization and consolidation have been generative.[26] A recent edited collection, *Digital Media Distribution: Portals, Platforms, Pipelines*, brings together case studies across a variety of contexts to understand the "digital transformation of media distribution."[27] Global mobile media studies scholars like Marc Steinberg have pointed to a power tilt toward distribution with the concomitant rise of mobile internet and digital platforms that require researchers to consider the circulation of content, discourse, and money as mutually constituted.[28]

Drawing on such works and extending them, I outline the specifics of distributing to mobile phones. There is a particularity to the mobile internet that needs to be differentiated from any general digitization. What happens when streaming portal and short video platform distributors target audiences who they know are accessing their content on mobile phones (and not on smart

TVs or laptops or desktop computers)? This kind of specificity is worth considering in India, where as a 2022 study by Kantar and the Internet and Mobile Association of India (IAMAI) indicated, 100 percent of active internet users access the internet (for entertainment and financial transactions) on their phones, but only 15 percent do so on personal computers (laptops/desktops), and 13 percent use other devices, including tablets and smart TVs.[29] The relationship between distribution infrastructures and audience/user imaginaries is crucial to understand, and this book explores several approaches—such as telecom bundling of streamers and the use of software development kits to extract phone data for credit scoring and digital lending—to distributing TV content and money to mobile phone audiences/users.

It is rare for a book to focus on the distribution of both content and money. This entwined focus on both fills a gap in distribution studies scholarship and demonstrates that digital transactions supported by mobile media continuously transform content and money into data and vice-versa. The mutual constitution of digital content and digital money has only deepened through the Indian government's sanction of a "cashless economy" since 2016. Through the demonetization (*notebandi*) drive, the Indian government encouraged digital transactions through mobile money services, thereby boosting the platform economy: mobile money made it possible, for example, to subscribe to video-on-demand streaming subscriptions in a country where the vast majority do not have a credit card.[30] "Digital monies," as Athique and Parthasarathi put it, "operate simultaneously as logistics, resources and commodities in the platform economy, interacting with and under-writing the network infrastructure."[31] This is evinced worldwide in the becoming social of digital payments with the accelerated convergence of social media platforms and digital money platforms: emojis/comments accompany Venmo transactions in the United States and followers bestow virtual gifts, bought with platform currencies (exchanged for renminbi/rupee), on their hosts in livestreaming apps such as Momo Live and Moj Live in China and India, respectively.[32] Platform capital and creator labor circulate as money becomes mediated/transformed into livestreamed virtual gift objects appearing in the form of pictographs/animations of bouquets and yachts, which carry with them the affections and expressions of followers as the laboring streamer interacts with their gifters.

India, this book claims, is a key site for the development of media distribution studies and platform theory. Whether it is the critique of India

Stack, a state-corporate initiative of pooling application programming interfaces (APIs) that contain applications for mobile phone data authentication and aggregation, or explanations about the initiatives of hyperlocal content delivery networks like picoNETS (vis-à-vis the usual stories about global CDNs like Akamai) to deliver faster video-on-demand services, the book's focus on mobile media distribution from the perspective of India (with comparative examples from other countries of the Global South) will provide readers with a better grasp of digital infrastructures in their historical and geographical diversity.

When the internet emerged, many thought cybercafes would be the path toward India's digital development, but that faith has long since been pinned on the mobile phone. Nimmi Rangaswamy suggests that the growth of cybercafes has shown mixed results in different parts of India.[33] However, mobile phones, being relatively more accessible, affordable, and easy to use than computers, promise the possibility of closing the digital divide through consumerist bottom-up interventions, even as questions regarding restricted computability remain.[34] Parallel to the shift from cybercafes to mobile phones, discourses around information and communication technologies for development (ICT4D) have morphed into mobile for development (M4D) drives.[35] But through all this, the one constant is the "access doctrine," which Daniel Greene has written about in the context of the United States. This doctrine makes "common sense" of a very contentious idea that the "problem of poverty can be solved through the provision of new technologies and technical skills."[36] I will think through this access doctrine and the "political economy of hope" in the context of so-called developing countries like India through a twin focus on media distribution technologies and aspirational politics.

Situated Understandings and Limits of Redistributing Access

Mobile phones, because of the scale of their uptake in emerging economies of the Global South and for being the "first access" for millions of people to electronically mediated communication, have been hailed as "transformative technologies" with "disruptive potentials."[37] That said, access to mobile phones cannot be assumed to automatically translate to empowerment. Mobile access is contextual for many women in India who might be using their father's or husband's phone for a few minutes a day.[38] Access to mobile phones for working-class Dalit women content creators has provided much-needed visibility and helped their content spread further, but such mobile access has also increased risks of surveillance and harassment. A blanket statement suggesting that mobile phones as technologies can transform society and culture "ignores the weight, social and cultural importance, and the politics of existing structures, as if they can be simply replaced."[39] In India, these societal structures include class, caste, gender, and religion among others.

The parentheses in "mobilities (un)limited" underscore the tension between the supposed boundless (or limitless) potentials of a mobile phone in India and the societal hierarchies and structural limitations that condition the contextual mobile phone practices of particular communities. With a concept like mobilities (un)limited, the book critiques hagiographies of mobile phones that are trumpeted by corporations that gain from phone subscriber expansions and instead focuses on local understandings of mobile media permeation and incorporation. This includes an engagement with mobile phones "within their culturally embedded everyday uses and settings," what Jo Tacchi calls "meaningful mobilities."[40]

The focus on mobile media distribution in conjunction with mobilities (un)limited in this book also helps bring out the discrepancies between the way phone users/audiences and use-case scenarios are imagined by distributors of content (streaming videos) and money (payments/loans) and the actual uses and meanings mobile devices acquire in situated contexts and layered environments. Lending platform tech enthusiasts I met in Bangalore and Mumbai waxed eloquent about the power of AI and alternative data points, where in cases of thin credit history files in India, loans were underwritten based on social media behavioral activity tracked on phones. This kind of purported financial inclusion through digital inclusion forgets about shared ownership of phones, where the SIM card might be in the wife's name but most of the day the phone is being used by the husband. The phone calls and contact list data may not be representative of the woman's activity. A lending company professional having worked in the Indian financial technology (fintech) sector for more than twelve years admitted to me during an interview that the loan underwriting models for women have to be very different if they are to truly reflect their phone-based social or financial behaviors. Many social activities remain beyond the purview of mobile phone usage, and the noise in the data evades capture by machine learning algorithms:

> If you really truly wanted to underwrite women digitally you have got to come with a very different underwriting model, concept, system because . . . the devices are very rarely used . . . so much that they [AI] depend on, data that only exists for a very small percentage of the population who live like this perfect life where everything is, you know, in their name, on their device . . . that is not the reality of this country . . . it is simply not true that you can get a perfect picture of somebody from a device. There is a lot of noise in that device data.[41]

Drawing on meaningful mobilities in order to theorize mobilities (un)-limited helps this book to discern the gaps and frictions between state-corporate distribution imaginaries that set developmental aspirational targets for new-to-mobile-phone consumer-citizens and what changes and futures are actually desired, aspired for, and enacted by community members with mobile phones based on their local knowledge systems. Local cultural values and meaning systems shape mobile phone usage practices that produce noisy data, thereby eluding algorithmic governance.

Here, I am neither suggesting that meaning only resides at the level of the user nor fetishizing the agency of the user compared to agencies of workers laboring at different points along the distribution supply chain. I most certainly disagree with people who find studies of distribution to not be meaningful. How can arranging things, directing flow, setting up blockages, and regulating access not have meaning? What I want to underscore is that along with an attention to media distribution backends, one also needs to appreciate the wide and deep background of users (their historical upbringing, their informal socialities) when they appear at the user–screen interface. After all, not only are the platform backends of a loan app opaque to a user, but the backends of social structures and cultural histories accompanying an individual user's life also might appear murky to the platform's machine learning algorithm.[42] Some data analysts I met acknowledged the flawed fundamental assumption on which the alternative scoring algorithms were based, which was that most people will become phone-enabled digital in their behavior. However, they reasoned that one had to become comfortable with that premise and work on how much can be known from that somewhat limited phone data activity so as to democratize credit, reach the needy, reduce customer acquisitions costs, and remove poverty. Faith and investment in phone data, AI-backed real-time algorithms, and behavioral finance to divine/discern reliable borrowers relied as much on their performative efficacy as on their functional precision.

From the Indian state to tech start-ups, and from policy wonks to digital evangelists, all see the mobile phone—equipped with mobile internet and apps—to be the technological solution to the "digital divide." This makes the mobile phone a critical material-discursive object in the rhetoric of aspirational politics, and here media distribution and aspirational mobilities are related, for the hope or the promise is that the mobile phone is key to any technological solution to solve the problem of access. Access to digitality (through the mobile phone) is equated to access to governmental welfare services, mobile payments, and crucial infotainment. Digital inclusion, or mobile inclusion, is equated to financial inclusion and, in some cases, even social inclusion. While these equations are certainly not as smooth as technocratic regimes might suggest, the mobile phone becomes a pivotal actor in the efforts to newly distribute or redistribute access. But redistributing access to services, money, rumors, and entertainment cannot just have instrumental ramifications of reshaping income and social inequalities; they will cause cultural clashes and affective ruptures as multiple publics and shadow economies respond and react to (im)mediated information, data, and money spread and leaks.

Such popular and populist responses afforded by mobile phones vary from progressive civic engagement to mob vigilantism. Information and rumor cut and spread across a variety of digital media platforms powered by the cell phone, thereby mobilizing sensing, affect, and multiple sites of attachment. Ravi Sundaram calls such phenomena "event chains" in a "distribution engine"—an affect machine and a crisis machine, dependent on the habitual microactions of likes and forwards.[43] Bishnupriya Ghosh observes that such "affective distributed politics" of viral intensities may not bring systemic change but signal toward a media populism that shapes the sensibilities of political subjects in India.[44] Contemporary India, after all, is youthful "millennials' India," home to the "stakes and struggles of a digitally savvy generation whose political imagination is inextricably linked to digital cultures as tools, habits, and sensibilities."[45]

Technological solutions to transform socioeconomic inequality continue to dominate in India. Some predatory microloan apps like Udhaar-Loan and FlipCash promise money to the "unbanked" (or "newly banked"), who have no credit histories, but the hidden rider is that they gain access to the user/customer's mobile phone data, including WhatsApp contacts.[46] As mobile payments get exchanged for WhatsApp socialities, distribution networks become places where money becomes data and data become money. These predatory loan apps are therefore not merely economic payment technologies, but as Lana Swartz notes, "their politics are a communication politics" of circulation.[47] When users cannot pay the money back, the loan apps, through their telecenters, get in touch with the customer's WhatsApp contacts (relatives and friends), humiliating the defaulters. In some of my conversations with such debt-ridden customers—whose microloans quadrupled over four weeks due to hidden interest rates—they have told me about the horrific experience of receiving threatening calls from the loan app companies. In these aspirational stories gone awry, new limits are drawn, stretched, and short-circuited.

Throughout the following chapters, I focus on this relationship between aspirational politics and mobile media distribution: how aspirations shape the socio-technical practices of distribution, as well as how distribution technologies create hopes and aspirations. Many business leaders and academic scholars believe that India has great potential but that not enough has been done to tap into it. Often India is compared to China as being the "other superpower"—and, like China, is one of the world's "indispensable emerging markets"—but does not have China's infrastructural might and manufacturing machinery.[48] This narrative is worth repeating as one asks whether mobile media technology will be *the* technology that will unlock the yet suppressed potential of India. Will mobilities (un)limited lead to "unlimited India"? It is not just the tech start-ups operating from Bengaluru and New Delhi that paint aspirational pictures around the mobile phone. Even the Indian state finds the mobile internet imaginary to be conducive to rebranding itself as an expeditious and innovative machinery (instead of a corrupt bureaucracy) and in prescribing specific aspirational targets to its citizens—who are also all (potential) mobile phone users.[49]

From Mobile Phone to Mobile Media

Darin Barney has observed that the valorization of mobility did not begin with the smartphone or the iPhone. The modernity thesis believed in achieving freedom as mobility through technology, and the mobile phone is a recent entrant in a host of former inventions in the area of transport and communications—including railways, cars (automobiles), radio, and television—that problematically promised freedom as movement.[50] This is a book about how the encounters of a diverse population of Indians with the mobile phone are part of the country's ever-shifting modernity project.

In India, the mobile phone reached the "masses" faster than either the car or radio or television and thus became a "mass media" very quickly.[51] According to Assa Doron and Robin Jeffrey, for India's disadvantaged people who could not reap the opportunities provided by the post offices, newspapers, or television studios—either because these readers/audiences were illiterate/disempowered or because the institutions were elitist/ exclusionary—the mobile phone offered the potential to change accepted (mal)practices, where every person has now become their own publicist. The authors found that seemingly unshakable strictures such as caste-based inequalities were undermined by mobile phones in India.[52]

Doron and Jeffrey's important book details the entanglement of mobile phones with business processes, agricultural practices, politics, and gender relations. However, most of their analysis is in the earlier period of the cell phone's ascendance in India—that is, from 1995 to 2010. Since then, as I argue in this book, several key developments have taken place in hardware (memory cards), messaging and video apps (WhatsApp and TikTok as well as apps made in India such as Moj and Josh), mobile payment and loan apps (Unified Payment Interface and mPokket), and cellular data services (Jio). These developments have led to such impactful changes in sound-and-image distribution, credit and cash flows, and "prosumption" processes in India that they merit a book-length study accounting for their scope and scale.

Arceneaux and Kavoori remark that mobile devices "are no longer limited to voice and text communication" and have become "quite literally portable multi-media centers."[53] As I will be discussing mobile phone applications and mobile internet services, streaming media, and financial technology infrastructures, I will attend to the intricacies of the "necessary convergence" of these manifold technologies and services into the erstwhile phone, which has transformed the mobile phone into mobile media.[54] Cara Wallis writes about the necessity of such a convergence of several uses in one mobile phone for lower-middle-class migrant women who come from rural areas to work in the cities of China and can only afford a mobile phone. Such migrant laborers need the mobile phone to serve many purposes.[55] Similarly, for most Indians, a mobile phone is the single digital device they possess, and the fact that it can be used for booking a train ticket, sharing songs, uploading a video, and making payments is a "necessary convergence" of functions and affordances.

While I do use the term "mobile media" for contemporary hyperconvergent mobile phones, I acknowledge Mara Mills' argument that particular histories (and futures) of the telephone, including audiovisual convergent proto-Skype technologies like picturephone/videophone, are elided in the shift of nomenclature toward "mobile multimedia." Also, "mobile media" need not be exclusively associated with mobile phones, because historically, a range of media have been mobile—as a writing and inscription technology, papyrus was a mobile medium, and various portable gramophones, mobile cinema vans carrying projection equipment," and radios could be considered mobile media.[56]

As a project about media distribution, this book has a responsibility not just to think of convergence in terms of media objects (film, television, mobile phone), but also to comprehend the impact of convergence on how distribution technologies have historically become contentiously and collaboratively entangled. Today, the streaming content that reaches the smart TV and smartphone is often not broadcast through a multicast signal and intercepted by a satellite dish antenna; rather, these are unique copies being pulled from data center networks by each request from digital devices. That said, local CDNs that distribute this streaming content today collaborate with local *cable wallas* (cable operators) who often still remain the last-mile operators that they once were in the heyday of satellite cable television. Now, they enact the role of internet service providers as last-mile operators.[57]

In this book about mobile phones and limits, I argue that memory cards pushed against the limits of data storage (especially vernacular music files) and that WhatsApp's viral mediations broke space-time barriers. Memory cards are key flash memory technologies that support storage and circulation of songs and music videos. In 2010–2015, there were many regions in India where the bandwidth to stream music or videos was not available, and people would go to "download vendors" who downloaded songs onto their memory cards (also referred to as microSD cards and colloquially called "chips"). These download vendors doing the "chip loading" often ran electronics stores or mobile repair and phone talk-time recharge shops, with laptops containing songs placed on the storefront and hanging placards advertising that songs could be loaded there (see figure I.1).

Figure I.1
Posters advertising Haryanvi, Mewati, and Pakistani mp4 songs for chip (memory card) loading in an electronics and phone store in Mewat (photo taken in Dec 2012). Photo taken by Abhigyan Singh.

The mobile phone has a slot where the memory card can be inserted, and then one can hear songs and watch music videos stored in it. With the continuing growth of cheap mobile phone handsets since the 2000s, the music economy in India had gradually shifted from CDs/DVDs to memory cards. From 2014 onward, the ascendance of memory cards has been tempered by the popularity of WhatsApp—an over-the-top (OTT) service that made sharing music videos easier—and the entry of cellular operator Reliance Jio in 2016, which made streaming services available to wider population segments. This "Jio effect" accelerated the concomitant rise of homegrown vernacular apps (like ShareChat, Moj, Gaana, DailyHunt, and Josh) and Chinese platforms (TikTok, Kwai) in India.

The smartphone—along with other mobile machines like the portable radio, iPod, train, and car—is an important heuristic device when considering the relations between transport and communication systems and connections between geographical movement and social mobility. Several chapters in this book deal with such relations: mobile phone users in the rural Mewat region desire motorbikes so they can travel to urban Delhi, and they use mobile phones to circulate locally produced vernacular Mewati music videos of themselves posing on bikes.

Along with this, the focus of the book regarding mobilities is not only concerned with the *movement of content* across entertainment apps and mobile media devices, what Chuck Tryon has called "platform mobility," but also with the *movement of money*, something that Lana Swartz has pointed to as an area to further research.[58] After all, "innovations in money and payments during the modern era have tracked alongside changes in the underlying communications infrastructure," whether it is paper checks and money orders sent through paper mail and express shipping, or digital payments through ATM cards, e-wallets, and loan apps.[59] Innovations in mobile payments have been devised in Global South nations. In 2007, Kenyans started using mobile phone company Safaricom's M-PESA service to send money to relatives through text messages. M-PESA achieved widespread popularity in Kenya much earlier than Venmo did in the United States, and since the 2016 demonetization drive, India has witnessed one of the fastest growths in mobile payments.[60]

This book is not just about the mobile phone (or mobile media) considered as a stand-alone object, or screen, or something to scroll and view apps on. This book is about mobile media distribution and hence tries to situate and relate the mobile phone apps to the multiple sites of intermediaries and infrastructures that make it possible for the app to function. It tracks encoding strategies and computational processes constructed by CDNs to, for example, make streaming a TV show possible on a mobile phone. It takes seriously the challenge of studying how processes of distribution change once people (like in India) start using the internet on their phones much more than on their computer browsers. Datafication works differently and more precisely in the case of mobile internet than web browsers. Jennifer Pybus and Mark Coté explain that users on the web are identified by cookies that can be deleted or obfuscated through circumvention technologies. Mobile applications rely not on cookies but on third-party software development kits (SDKs), which are far more embedded inside the source code of these applications.[61]

This has wide-ranging implications for mobile data distribution, particularly regarding how sensitive user data is stored, exchanged, and moved during monetary transactions as part of digital payments, lending, and insurance. Loan apps, for example, rely on third-party SDKs to check customer identification and credit scores before providing them loans. But what if these third parties siphon off customer data, as is the case in some examples I discuss in this book? Mobile phones are considered to be the device for the financial and social inclusion of the unbanked poor, but what are the costs of such inclusion if the mobile internet ecosystem by design allows for data leakages and permissiveness? Mobile internet structurally asks for a recalibration of the limits of data mobility, privacy, and promiscuity.

During the late 1990s, India's telecom revolution emerged as a signifier of global neoliberal politics and techno-nationalism. Even as the postcolonial Indian nation-state has over the years increasingly privatized the telecom sector, it continues to see itself as a network builder and champions telecom revolutions for both national and international recognition.[62] While the conceptualization of mobilities (un)limited should be related to macronarratives of India's development, it is crucial to consider the Indian growth story beyond the upper echelons of policymaking and the statistical models of economics. The future potential of India is, after all, about the human development of 1.4 billion Indians, whose capacity to aspire has been fueled by neoliberal globalization. Mobilities (un)limited is very much about aspirational mobilities.

Aspirational Politics

"Aspiration" is a buzzword not just in electoral politics but also in the politics of the everyday in India. Finance ministers pronouncing budget policies talk about "aspirational India" for the "neo–middle classes," an emerging group that has broken the shackles of poverty but has not yet stabilized into the middle class. In this case, it is not income but the capacity to aspire that is perceived to be essential to being counted as middle class.[63] Political scientists have observed that Prime Minister Narendra Modi and his ruling party, the BJP, have rhetorically deployed the term "neo-middle" in several elections as a way to invite "lower class aspirers to join the ranks of traditional middle-class identifiers."[64]

Some Indians aspire to status in their caste-religious community, some aim for socioeconomic mobility, some want to move from one region of India to another region, and still others want India to be counted among "strong" nations in the world. This book investigates the connections between such aspirations and the mobile phone.

People from small towns in India migrated to big cities to become construction workers, maids, call center professionals, software engineers, and security guards, and while some of them have remained within the lower-middle-class income bracket, they have continued to aspire to someday becoming middle-class and even upper-middle-class people. Aspirational mobility is therefore about both (increased) spatial and (upward) social mobility. However, economic downturns like the COVID-19 pandemic have led to restricted mobilities and downward mobilities where vulnerable neo–middle-class aspirers lost jobs and fell back into poverty or had to reverse-migrate back to their villages from cities as strict lockdowns were enforced in urban areas. Aspirations entail hopes and risks.

Anthropologist Arjun Appadurai argued that the upper middle classes generally tend to have a higher "capacity to aspire" than low-income groups and subaltern populations because they have a "wide range of ends and means" and more opportunities to try out material goods and the experiences such commodities afford.[65] Low-income groups and many marginalized communities in contemporary India have a greater capacity to aspire than they did before, not because their incomes have significantly increased, but due to media exposure and their awareness of what others are doing. Ellen Barry notes that mobile phones have furthered the production of aspiration by bringing rural Indians into virtual, yet intimate, contact with urban middle-class lifestyles: "Indians are moving to cities in huge numbers; but even those who remain in villages, surfing the Internet on cheap smartphones, are drawing closer to the way city-dwellers think." The neo–middle class is no longer satisfied with rationed welfare benefits and daily necessities but desires a consumption-oriented lifestyle.[66] Memory cards, WhatsApp, Jio, and TikTok are accelerating the ability of people to know about what others are doing by allowing for greater socialization of desires and lifestyles. Therefore, aspiration is not necessarily an individual-level phenomenon but a relational one, something that Appadurai has stressed: "[Aspirations are] never simply individual (as the language of wants and choices inclines us to think). They are always formed in interaction and in the thick of social life."[67]

Appadurai's theorizations on aspirations also link cultural capacities to potential economic mobilities and therefore provide a "relationship between culture, poverty, and development"—something that has been emphasized by Nimmi Rangaswamy and Payal Arora.[68] These researchers discuss some of the ways in which the aspirations of poor youth living in the slums of Chennai and Hyderabad in South India manifest in their everyday mobile internet use on their modest-feature phones. While accessing social media sites like Facebook, these youths sought "interactions with people from an elevated social status to fulfil aspirations of social mobility," which included "heterosexual dating possibilities aligned with possibilities of upward class mobility via friending women from higher social segments"[69]

I conceptualize mobilities (un)limited by connecting aspirational mobilities to geographical movements. In doing so, I focus on the intersections of mobile media studies and the "new mobilities paradigm," with particular attention to the simultaneous occurrence of transregional mobilities and deepening regional meanings and affiliations in India.[70] For example, Bhojpuri music enthusiasts among the "lower-caste" youth walk through the streets of the city Gaya with hopeful strides as they listen to songs on their portable Coke-bottle-shaped audio players, which have memory card sockets. Such mobile listening practices signify to the world that they have "finally arrived," breaking the shackles of caste even as they are rebuked by upper-caste men who mark such practices of movement (which are also media practices) as "tasteless cultures." Gaya is located in the state of Bihar, a region of India known for its caste fault lines (and mobilizations).

Aspirations are not confined to media consumption practices. As in many other countries, India is witnessing the "platformization of cultural production" through smartphones.[71] The platform economy has created platform labor, with a new group of creators using apps like TikTok, Moj, and BigoLive to reach fans and followers. Sathana Chinnadurai, a housewife in the remote Tamil (South Indian) village of Nachelura, joined TikTok in 2018 and sang and danced to songs from Kollywood films (Tamil-language cinema and media, with "-wood" suggesting an extension from Bollywood and Hollywood). She amassed 700,000 followers and was invited to do live performances in cities like Madurai.[72]

Chinnadurai's success story has made still-struggling content creators hope that someday they will also amass a great number of followers and receive financially lucrative invitations to perform onstage. While Chinnadurai's example is a story of aspiration being fulfilled, many other social media content creators are yet to be rewarded, as the reward system of platforms like TikTok and YouTube is highly uneven. That success is so heavily based on quantitative metrics suggests limitations to the "aspirational labor" such platforms try to foster.[73] Furthermore, Chinnadurai has faced abusive trolls, including a fake online obituary circulating the untrue news of her death, thereby encountering the limits of pushing societal boundaries through platforms. These findings about the ambivalent sociopolitical outcomes of the introduction of mobile phones can be seen in other countries. Examining the link between the eager uptake of mobile phones and refashioned forms of aspiration in Mozambique, Julie Soleil Archambault explains how "mobile phone use simultaneously builds on socioeconomic hierarchies while also playing a decisive role in the negotiation of socioeconomic disparity."[74]

The rise of Digital India has also accelerated cybercrime in the country, and in some cases, opinion is divided as to whether aspirational youths' mobile media cultures should be considered innovative entrepreneurship or cybercrime. Let me recount a particular case study in the Indian state of Jharkhand. Rural unemployed youth of Jharkhand's Jamtara district impersonate bank officials and call customers in cities, asking for their card details and explaining that this information is required to offer the valuable customers particular gifts. Such credit card scams based on mobile phishing (vishing) have become rampant, made possible by the construction of 3G (and, later, 4G) cell towers in Jamtara and other rural areas of India.

During my fieldwork in the village of Karmatand in the Jamtara district in July 2024, a village known for mobile phishing operations, I spotted a cluster of antennas on four cellular towers, a tele-density that seemed disproportionately high for a small village (see figure I.2). The digitization of India's economic periphery (through cell phones and cell towers) has not only increased the capacity to aspire but also accelerated the rates of cybercrime. Growing up in the Jamtara district, where there is massive unemployment, many young people had their education shortened, as they were compelled to leave school to work and earn money—circumstances that make cybercrime an attractive option. Some scammers have used the money from these phishing operations to construct houses and buy motorbikes. In a freshly painted wall ad, I saw the telecom company Airtel announcing a new cell tower in Karmatand. Alongside ads for telecom networks and cell phones available on loan, I also found adverts for cement companies, indicating a construction boom in the village. Perhaps these corporations also tacitly acknowledge the entrepreneurial dimension of mobile phones that are emerging in Karmatand.

Figure I.2
Four cell towers in Karmatand Village in Jamtara District. Photo taken by author.

Based on my interviews with locals in Jamtara, I found that some regard these youths as aspiring to be entrepreneurs through this phishing activity, while others consider them as engaging in a new kind of crime. Mobile media devices generate, or help realize, limitless/unbounded aspirations, although chasing them could lead to severe challenges and risks on the way. The risks undertaken by the youth in Jamtara while conducting these phishing scams is something that needs to be taken into account as a limitation of such aspirational mobilities. Some bystanders in a neighboring village referred to the phishing operations as *dhanda* (business) and remarked that those who engage in such *dhanda* do not leave it soon because even though it is risky, the rewards from a successful phishing operation, if it works out, are quite high.

There are limitations to aspirational mobilities because they are managed by neoliberal forms of governance initiated by the state and corporations. It is not enough to study aspiration as individual ambition and hope, because states and corporations also actively participate in prescribing particular aspirational targets for consumer-citizens. With the neoliberalization of the Indian economy since the early 1990s, the state and nonstate actors have sought to "create heightened aspirations and expectations, promoting the ideology of self-making."[75] I argue that the conjunction of mobile phone apps with the platform economy is a continuation (and in some cases, intensification) of these ideologies of self-governed individuals, where the mobile phone is presented as one of the self-development tools that purportedly help individuals self-propel themselves as successful entrepreneurs. Prime Minister Narendra Modi's Digital India campaign had several ambitious targets, not just for India as a whole but for Indian citizens, including opening bank accounts and using online cashless transactions, in which mobile phones would play a crucial role (see figure I.3). The start-up app companies also present a future India for the country's "app-izens," as if apps will solve all their problems, including unemployment, through innovative entrepreneurship. National development is explicitly related to personal aspirations in these governmental and corporate messages.

Pradhan Mantri MUDRA Yojana (Prime Minister's Micro-Unit Development and Refinancing Scheme, PMMY) is a micro-credit/low interest loan

Figure I.3
Opening bank accounts through phone selfies (Union Bank's mobile banking ad, Mumbai). Photo taken by author.

program to incentivize the participation of neo-middle classes, especially women, in the market and encourage them to be aspirational entrepreneurs. In addition to some government banks, private sector companies including gig economy players have been allowed to provide loans to their contractors. According to Priya Chacko, initiatives such as PMMY help the BJP government to remake neoliberal self-improvement projects within the cultural framework (patriarchal ideologies) of Hindu Nationalism where the patriarch (state and leader) is shown to be taking care of and empowering self-disciplined and responsible women entrepreneurs/borrowers. PMMY promotional campaigns heavily feature Prime Minister Modi along with young women entrepreneurs.[76] Aspirational politics becomes interweaved with social policies in new contests over the terms of inclusion and exclusion that inaugurate novel relations of credit, debt, nationalist ideologies, and market dependencies. I study such financialization processes in the context of mobile money in this book. The Sanskrit/Hindi term for "aspiration" is "Aakanksha"/ "आकांक्षा" which connotes expectancy or yearning that arises from the need for sense-completion. Aakanksha is the chosen brand name for promotion of local products made by grassroots entrepreneurs in "Aspirational Blocks," those districts/blocks identified by the Planning Commission (*Niti Aayog*) for rapid development.[77]

There remains a vast population in India who are neither middle class nor working-class, but rather have faced long-term unemployment, relegated as surplus population or "lumpen/proletariat." Many among India's so-called demographic dividend, the youth, are at the edge of falling into being such "dangerous classes" who might challenge the state with (un)civil protests or acts of criminality.[78] The state hopes to bring them into the fold of neo-middle classes through schemes such as Digital India, Skill India, and Make in India, with the call to aspire doing a lot of affective and emotional work. The state government of Madhya Pradesh went so far as to say that the category of "unemployed youth" should be changed to "aspirational youth."[79] Such renaming tries to hide the job crisis but ends up exposing the empty promise of aspiration. Several of my interlocutors have challenged the aspirational framing of both academic theorists and policy-makers arguing that such an optic might end up reifying state-corporate discourse.

The official modalities of ambition and development encoded in "aspiration" as a state and corporate project must be differentiated from the individual and collective desires of citizens. Aspirations, as Peter van der Veer discerns, might even compete in an unequal power field, where the visions of planners might be very different from the "perspectives of those who are living in these visions."[80] Aspirational politics remains open to various interpretations, motivations, and experiences.

The aspirations stoked by corporate talent scouts as part of recruiting and nurturing talent for short video platforms may not always neatly overlap with the desires and ambitions of content creators. One particular influencer from Kanpur who I interviewed noted that she did not like being constantly asked to collaborate with Bollywood celebrities unless there was an artistic merit to such collaborations. Often these collaborations were presented to her by talent agents as opportunities to network and increase follower counts, but she believed follower counts will be meaningful only if she is recognized as an artist, and that will happen when she can devote time to her craft of writing the scenarios she enacts in her short videos rather than spending that time socializing or networking.[81] A microinfluencer in rural Rajasthan told me the short video platforms tacitly pressure creators to paint a wholesome picture of rural India, not its problems, which he refuses to do. There are thus these ruptures between corporate discourses of aspirations and networking and creators' own ambitions about the kind of content they want to make that rejects neoliberal determinisms. It is in these slippages between state-corporate discourses and individual desires that I conceptualize "aspiration politics" in this book.

Mobile Media Distribution

Aspirational politics is connected to media distribution because—with a wider set of offline publics in India becoming part of online networked public spheres—interactive possibilities have expanded. Even though regional disparities remain, people from different regions are getting to know one another and meeting up online on various apps and platforms through mobile media. Through exchange and circulation of audiovisual content, they are learning about each other's lifestyles and livelihoods. These apps and platforms are not only spaces of gathering but also conduits of distribution. Whether it is music videos or cashless digital money, they arrive on mobile media devices through various digital intermediaries and an assemblage of platforms and infrastructures.

Sometimes, an act of mobile media distribution might involve the informal exchange of memory cards containing downloaded music videos, which can then be inserted into mobile phones and watched. At other times, distribution involves the formal media industries and corporate power, which controls the flow of streaming content into mobile phones, limiting what media is available where and for how long. Donoghue et al. note that researchers should study not only the macrolevel political economies of oligopolistic media distribution mega-corporations but also the "micro-social practices of individuals" who work at various stages of the distribution supply chain. This helps in examining how corporate power and control are "devolved, divided, and differentiated, that is to say, distributed" among these different distribution professionals.[82]

In several chapters of this book, I interview professionals involved with the distribution backend at various stages—content curators, catalog experts, content delivery networks, encoders, cloud architects—who enact socio-technical practices by working with transcoding algorithms, edge computing technologies, and data centers to ensure that streaming content reaches mobile phones. Here, I espouse Joshua Braun's call to think of "media distribution" as a "collection of socio-technical practices."[83]

In what ways are aspirational politics connected to the socio-technical practices involved in media distribution? In this book, I examine how the cultural politics of aspiration shapes and is shaped by the new mobile/digital distribution technologies. The distributors of content sometimes tend to think they have perfectly conjured up the aspiring neomobile audience—what such audiences want to see and hear and in what conditions are they watching the content—by reading popular sentiment and by designing use-case scenarios. However, the distributor's assumptions could be wrong, and there are limits to their research and intuition. Furthermore, I study both the impact of media distribution and how the distribution takes place. Here, I particularly focus on the role of digital intermediaries and informal operators as well as the particular affordances of specific apps in the way they enable or constrain interaction, communication, and distribution.

I will now discuss each of these above-mentioned aspects of mobile media distribution ecologies, but let me begin with the connection between aspirational politics and media distribution. To offer an example, the Indian state, especially in the decade of 2010–2020, which I study in this book, has time and again created an aspirational narrative of financial inclusion for the unbanked millions in India. This is not merely a rhetorical discourse: It has been enacted through an Aadhaar database of biometric identification details. This database not only helps to authenticate users but also provides opportunities/access to third-party entities (including private businesses and for-profit and nonprofit start-ups) to build digital services (apps) with some regulatory safeguards. Aadhaar is therefore both a digital storage technology (database of 1.4 billion users—that is, India's population) and a digital distribution technology. Added to the Aadhaar database is India Stack, a set of cloud-based application programming interfaces (APIs) built atop the database, which makes it possible for private companies to verify identities and offer financial services.[84]

The Stack's APIs allow it to monitor the data transactions of the as yet unbanked millions and then calculate their creditworthiness in the absence of other kinds of formal financial data. Understanding this storage and distribution technology, which is so closely connected to the mobile devices of India's population, is crucial because this kind of data capitalism is not essentially being run by Big Tech companies extracting data trails based on internet browsing history. Here, the (Indian) state actively provides digital Aadhaar ID details to fintech companies in the name of financial inclusion. In return for services, Indian welfare recipients therefore relinquish their identification details to the state, which are then shared with fintech companies for data analytics.[85] This movement of data and financial services across the different layers of the India Stack is important to examine in the various case studies concerning credit card frauds and microloan app scams that I study in this book. The India Stack is therefore a key digital intermediary that connects the state, corporations, and citizens in India, turning "distribution infrastructures into analytics engine."[86] The mobilities across the different layers of the India Stack suggest another kind of unprecedented interaction among the data-service-money distribution channels, an architecture and phenomenon, that is captured by the concept of mobilities (un)limited.

Another key aspect of studying media distribution has been to understand the role of "informal operators" who work at the edges of formal media industries or create their own alternative distribution networks.[87] The various vernacular music video distribution networks that I discuss in this book, which are enabled by memory cards that are attached to mobile phones, are based on the work of several such informal operators. They include "download vendors" who download songs from their laptops into the customers' chips, or memory cards. Similar activities are also performed by mobile stores or mobile repair shops, where the crucial format conversions are undertaken, so that CD/DVD files can be played on memory cards through mobile phones. These informal operators therefore act as pivotal nodes in the circulatory assemblages for music videos.

The case studies I discuss about memory cards are from the earlier period of the decade (2010–2015), and since then, some scholars have argued that there has been a formalization of media practices in India and elsewhere. With increasing formalization, digital intermediaries like search engines and social media have played a key role in news and rumor circulation. That said, digital intermediaries like YouTube (several news channels have their own YouTube channel) also require a whole group of formal and informal people and organizations such as talent agents and multichannel networks (MCNs) who extend strong influence over content creators and streaming platforms.[88] Based on my interviews with such scouts working for TikTok and livestreaming apps like BigoLive—and, later, for Indian short video platforms like Moj and Josh—I argue that they should also be considered to be part of the production and distribution ecology of streaming video apps in India.

Intermediaries are crucial to mobile payment transfers as well. Although the rhetoric of financial inclusion through the mobile/digital revolution often presumes to paint all human intermediaries as middlemen, brokers, and (informal) predatory lenders, the mobile money distribution systems in India (and various locations at the margins of the Global South) have their own set of human intermediaries. Ishita Ghosh and Jacki O'Neill list a number of intermediaries such as mobile money agents, cash collectors, and telecom service shops involved in telecom operator Airtel's digital wallet service, Airtel Money.[89] When it comes to the distribution of money through an online lending app or a mobile payment app, a whole host of intermediaries such as alternative credit scorers and identity-verification checkers—who supply these financial services, also called fintech as a service (FaaS)—are found to be embedded in the lending system.

The various intermediaries involved in the lending platform ecosystem and fintech infrastructure can be found to gather in the fintech conclave. I attended the Fintech Festival India held in early March 2024 at the Yashobhoomi Convention Center in Dwarka, New Delhi (figure I.4). Here, I found

Figure I.4
Fintech Festival in Dwarka, Delhi (March 2024). Photo taken by author.

stalls set up by loan app companies such as Fibe who were advertising that they cater to the aspirational youth, as well as fintech as a service (FaaS) players such as Mobilewalla, who promised AI-driven application programming interfaces (APIs) and SDKs to approve thin-file customers and thereby enable financial inclusion.

The statements they had on their pitch deck dashboards could be dismissed as glib corporate talk, but at the same time, there were eager professionals representing third-party fintech developers who seemed keen to take me through a flowchart of how they conduct face match and liveliness check as part of their offered suite of instant verification APIs for onboarding and digital know-your-customer (KYC) checks. The pictures of those who had their Aadhaar (unique ID) cards made early had to be matched with their recent pictures, and this required a "live" selfie to be taken during the loan app onboarding process. More nuanced algorithms were needed to make sure this was the same face as that of the person on the Aadhaar card. As much as this sounds like a techno-utopia, it was also a kind of normalization of digital surveillance, all done in the name of security and verification. These acts of data extraction as part of the loan onboarding process are characteristic of both legitimate and predatory loan apps, so it becomes very difficult to distinguish between the two, given the similarity of "dataveillance" practices involved.

In the several case studies in this book, the reader will encounter intermediaries involved in the mobile money distribution system in the form of cybercriminals or enterprising mobile phishers who rent other people's Jan Dhan accounts as well as telecenter operators who work with predatory loan app companies calling nonpaying customers.

Digital content media distribution does not depend only on circulatory assemblages and infrastructural specifics. It is also shaped by how user practices get entangled with particular affordances of platforms. André Brock lists the ergonomic aspects of smartphones, including their small size and easy handheld camera recording technologies, that have facilitated the growth of audio-video apps that allow for sharing activities such as "impromptu dances, risqué behavior, and moments of hilarity (and violence) that couldn't take place in more proscribed places."[90] The mobile phone equipped with sharing and forwarding apps puts content in motion by "lubricating redistribution networks."[91] Both WhatsApp and TikTok have been extremely popular circulation platforms, but their distribution capacities influence and are influenced by different user practices. In the case of WhatsApp, message forwarding through semianonymous WhatsApp groups leads to virality. With the TikTok app, the duet features and lip-synching practices contribute to mimesis and replication emerging with "imitation publics."[92] WhatsApp and TikTok intensify the blurring of boundaries of media "production, circulation, and use" because their user functions of sharing and forwarding are acts of (re)distribution.[93]

Digital India: Neomobiles, Data, and Platform Ecosystems

In June of 2018, I took a picture of an ad from Indian cellular operator Airtel promising "unlimited incoming and 3GB data" in Delhi (see figure I.5). Airtel is Jio's primary competitor, and responding to Jio's free calls, Airtel had to make incoming calls free too. A veteran tech journalist covering Indian telcos told me that Jio made calls irrelevant because they just had to be free. What mattered now was how much data a telecom service could

Figure I.5
Airtel ad promising unlimited incoming calls. Photo taken by author.

provide customers. Jio and Airtel started a data war contending for mobile phone users. This shift from call to data with the concomitant rise of 4G smartphones has made mobile phones the digital device for realizing the dream of Digital India—which at some level is about connecting 1.3 billion citizens even as digital inequalities remain patterned on societal inequities and differential access to education and health care.[94]

It is helpful to consider the range of differentiated uses of mobile technologies and access to various levels of mobile phones, data packs, and internet bandwidths. As Rashmi M. notes, "'Feature phones' have not completely disappeared after the arrival of smartphones," and there is a "large and growing market backed by increasing demand for used smartphones."[95] The mobile phone stores and mobile repair shops along the street, in many places and communities, carry both new and secondhand smartphones. It is even possible that transactions related to secondhand phones conducted with cash are not formally recorded. Given the kinship structure in the country, some secondhand phones are passed from family elders to younger kin once a newer phone is purchased. Amid this cultural economy of phone circulation and distribution, the Chinese company Xiaomi and the South Korean company Samsung have often competed with one another for the large Indian market, making a range of phones, including feature phones and budget smartphones.[96]

In an effort to find the "product market fit," Sajith Pai, a consultant to venture capitalists funding start-ups, created various tiers of phone app users (in "India 1," "India 2," "India 3," and "India 4") to suggest that different apps might be more suited for different tiers of users living in different categories of cities.[97] These market categorizations of phone and app consumers reflect the digital/mobile inequalities in India. Mappings and segmentations of users do not end at connections between types of phones and kinds of apps: They move further to include taste cultures. A short video platform content director and talent recruiter, Raghvendra Prajapati, told me his app caters to tier 3 audiences who use a cheaper Vivo or Redmi handset and are devotees of the Bhojpuri singer Khesari Lal Yadav. He emphasized that the content creators and influencers on his app need to know that they are creating for fans of Khesari Lal and not for the admirers of the Korean band BTS, who live in tier 1 Indian cities like Bengaluru. A BTS fan using an iPhone or Samsung smartphone in a tier-1 city would be totally unaware of Khesari Lal, and a Khesari Lal admirer in Gaya is most likely not going to know about BTS.[98]

Prajapati works for short video platform company, Moj, which like TikTok before, has been trying to challenge the hegemony of Facebook, Instagram, and YouTube, particularly while targeting regional and vernacular users/audiences in tier 2 and tier 3 Indian cities. While advertisers still stayed away from these Indian and Chinese apps, maintaining that they catered to wallet-thin lower-tiered consumers, venture capitalists did invest in them, hoping that these apps were soon going to have an enviable customer base as a whole aspirational neo–middle class in India (estimated to be 200 million people) acquired mobile phones for the very first time in their lives (termed "neomobiles").

Sharechat developers discerned that English-only BuzzFeed jokes and *Game of Thrones* memes were incomprehensible and rather estranging to their neomobile users (often referred to as regional or provincial or vernacular users), so they boldly decided to remove English as a primary language of content from their app.[99] Beyond making infrastructural and technical adjustments, I argue that ShareChat and other vernacular apps responded to a de-anglicized India (termed *Bharat*) within India, tapping into the aspirational politics of Digital India that shapes and is shaped by a new kind of Indian political subject. For these users, some would say that access and discovery issues are far more important than privacy and security concerns. In July 2020, the Indian government, citing threats to data security and data privacy, banned Chinese apps, including TikTok and Kwai. Thousands of Indian content creators who were enabled by these platforms, lost their followers and earnings. Several tech journalists called the Indian government's move to ban TikTok and BigoLive an assault on aspirations, especially those of the millions who were coming onto the internet for the first time. In an interview with me, one journalist noted, "Their [new mobile users and influencers] creative freedoms were being quelled when they banned TikTok."[100] Many credited TikTok with nurturing a kind of talent and expression that had just reached a certain scale, with India poised to witness the next wave of influencers rising up in its small towns when the apps were blocked.

The aspirational stories of these neomobiles, especially in economically backward states, certainly add a democratic shine to Digital India, but political economy studies point out the pitfalls of falling for the heady language of the campaign. As Pradip Thomas puts it, "The much-lauded growth in Internet penetration rates, the distribution of mobile phones, and digital access belies a real consideration of the fact that the epicenters of India's knowledge economy are state and city specific."[101] People in states such as Jharkhand and Madhya Pradesh (which have historically lagged behind in socioeconomic growth indicators) are today hailing Uber for rides and having goods delivered through Amazon. However, Uber and Amazon are opening their new centers in the already developed states of Karnataka and Telangana.

Prime Minister Narendra Modi launched the Digital India initiative on July 1, 2015 with the aim of connecting 1.2 billion Indians. Digital India was conceived as a public–private partnership (PPP), and several private firms were involved in shaping India's digital future. Telecom companies constructed new cell towers and linked them through fiber-optic cables.[102] The Digital India vision is not just about corporate players fulfilling the government of India's visions. The government has itself actively sought software technocrats to enroll more than 1.2 billion Indians in its biometric digital identity program, Aadhaar, and through the goods and services tax, it brought more than 10 million businesses onto a common digital platform. Soon after the launch of the Digital India initiative, the National Payments Corporation of India (with the support of the Reserve Bank of India and the Indian Banks' Association) set up a range of digital payment platforms accessible to citizens through mobile phone apps, including BHIM (Bharat Interface for Money) and UPI (Unified Payments Interface).

In creating some of these software-based storage and distribution infrastructures, particularly the India Stack, through PPPs, the role of iSPIRT (Indian Software Products Industry Roundtable) in consulting the government and interfacing between the government and software start-ups has been both crucial and controversial. iSPIRT started as a think tank in 2013, and its volunteers helped Indian software product companies develop. Soon after, iSPIRT's primary role morphed into positioning Aadhaar with product companies as a way to target Indian consumers (mostly through phone apps). This led iSPIRT volunteers to develop and promote India Stack APIs that allowed software companies to plug and play with the Aadhaar database.[103] As one journalist put it, "India Stack is the gate to Aadhaar . . . the gate to India Stack is iSPIRT": Any third-party software company that wants to use the Aadhaar database or UPI to provide/deliver financial or health services to Indian consumer citizens needs to curry favor with iSPIRT.[104]

As much as the government is regulating platforms, it is also creating new platforms. Adrian Athique noted that the Indian government is engaged in not only the *governance of platforms* (setting regulatory protocols for platform companies to adhere to) but also the *platformization of governance.*[105] The platformization of governance entails the government (in partnership with private players) designing entire suites of apps for last-mile delivery of various government services. This allows the government to serve, surveil, and interpellate citizens, to shape them as political subjects. For many, the face of Aadhaar, the technocrat Nandan Nilekani, summed up the Indian state's e-governance role by stating, "The fundamental nature of a government is a platform."[106]

Media studies scholars have argued that emerging societies are undergoing platformization across public and private sectors. Socioeconomic relations are increasingly channeled by online platform ecosystems that are fueled by data and conditioned by algorithms.[107] Media practices like music sharing and video distribution are now packaged and aggregated as data and content by algorithms, which are able to create value for these products and "make them legible as transactions with a digital footprint."[108]

India has witnessed an intense competition between Silicon Valley–based platform companies, Chinese video-streaming companies (until they were banned in June 2020), and Indian legacy companies that have made the digital transition. The OTT video platforms—from YouTube to TikTok, or from Amazon Prime Video to Netflix, or from Hotstar to Voot—have been fighting for the attention of more than 460 million Indian internet users, a market second only to that of China in size.[109]

Roadmap of the Book

This book is the story of the mobile phone in India from 2010 to 2024. With smartphones becoming more common in the second half of that time period, phone apps have become the "front doors" for corporations to reach potential customers and distribution end points for many governmental welfare services. These apps are also sites of user behavior data extraction. Mobile media distributors repeatedly told me that controlling different stages of the distribution supply chain is vital because that helps to collect data for targeting customers in even more personalized and customized ways. Mobile media are also the tools through which erstwhile marginalized communities are expressing themselves and making their presence felt in the public cultures of India. Mobile media distribution signals a shift from ICT4D to M4D discourses, with the "appropriate technology" of ubiquitous communication and service delivery for the vast global majority being the mobile phone (and no longer the computer). That said, by allowing for both state–corporate service delivery and the spread of user-generated expressive and relational content, like ICT4D, M4D tries to do away with the binaries of "top-down technocratic development policy" and "bottom-up community activism."[110]

Moving through the book chapters, I track the shifting usages of memory cards and WhatsApp in the early half of this decade (2010–2015) and the emergence of short-video platforms and loan apps in the latter half of the decade (2015–2020). This book traces the move toward increased platformization and formalization of cultural production, consumption, and distribution along with accelerated financialization in India. Each chapter carefully considers specific discourses of aspirational politics in relation to the distribution practices of particular mobile media ecologies.

The first chapter, "Loading: Memory Cards," concerns itself with the distribution assemblages of vernacular music videos based on the popularity of memory cards as storage devices that can be inserted into a mobile phone to watch the videos. The second chapter, "Streaming: (Un)bundling and Deep-Edge Computing," examines how demand for streamed content is sustained and accelerated by a crowded space of global and local streamers in India, such as Netflix, Amazon Prime Video, SonyLIV, Disney+Hotstar, and ALTBalaji, which offer video-on-demand (VoD) services through a variety of subscription-based and advertising-based business models. Based on interviews with people working at different stages of the streaming distribution ecosystem, I describe the work of intermediaries (involved in encoding, decoding, storage, distribution, and curating) who make it possible for the mobile phone publics to access cultural content. Audience imaginaries shape the work of these distributors, and there is an emerging discourse in the streaming industry to bring relevant content closer to the user—that is, a movement away from the center (data center) to the deep edge (cached storage close to user). This is a movement from cloud computing to deep-edge (mist) computing.

We see different kinds of aspirational politics in these two chapters. In chapter 1, I discuss the aspirations of youth belonging to particular regional vernacular communities (Meos, Bhojpuri enthusiasts). In chapter 2, I examine how streaming platforms aggregated as part of bundled offers from telecom companies are catering to the aspirations of Indian mobile phone users who are enjoying the possibilities of individualized and fragmented viewing on their newly acquired mobile phones (instead of shared family television sets) in a sustained manner for the first time. From the first to the second chapter, there is a shift from "informal" to "formal" distribution logistics, as well as transitions from "download cultures" to "streaming cultures" of viewing. This reconfigures discourses about piracy, consumer sovereignty, and data–energy relations.

The third chapter, "Spreading: Viral WhatsApp and Post-TikTok Futures," examines the particular affordances of WhatsApp and TikTok that enable messages to go viral in different ways. Deploying both ethnographic interviews and app walkthrough methodologies, I investigate app features and the hyperlocal and individual experience of such apps by new phone users (neomobiles). The seemingly boundless viral potentials of apps like

WhatsApp and TikTok make manifest the mobilities (un)limited discourse that on earlier occasions has been associated with memory card–circulated music and Jio's data plans.

When TikTok exited India, homegrown short video platforms such as Moj and Josh endeavored to recreate the app's success among new smartphone users in regional tier 2 and tier 3 Indian towns where YouTube and Instagram Reels had not yet seized the market. These short video apps replicated the TikTok interface, invested in infrastructure in efforts to match TikTok's machine learning–enabled recommendation system, and employed talent scouts to recruit local creators for their platforms. Both Josh and Moj conduct retreats and workshops for talented creators to meet each other and Bollywood celebrity mentors, fostering collaboration and aspiration. Earlier initiatives to tap talent for reality TV shows in India had experimented with renting out space in shopping malls or cinema halls in tier 1 and tier 2 cities, but these efforts did not extend to tier 3 cities, where there were no such halls and malls. Josh's innovative outreach efforts included sending mobile stages on canter vehicles to create spaces for talented villagers to audition. This is part of wooing neomobile regional users in India who communicate in non-English "vernacular" Indian languages.

If the first three chapters grapple with the circulation of audiovisual content, in the fourth and fifth chapters, I trace the circulation of money, credit, and cash. Mobile payment is a key distribution technology that developed initially in the form of mobile wallets, and then later, in 2016, with the introduction of UPI, Big Tech dived into this sector. WhatsApp also entered the mobile money market in India with the WhatsApp Pay feature.[111] Google launched GooglePay in 2017. Amid this mobile payment landscape, I examine case studies of credit card fraud in chapter 4, "Phishing: Insurgent Trajectories," and of predatory microloan apps in chapter 5, "Lending: Loan App Scams and FinTech Services."

Rural youth in Jamtara pose as financial service providers and call urban card holders, mimicking call center tones and rhythms and asking for credit card details. These impersonation tactics of Jamtara scammers create a recognizable space of proximate interaction in the minds of their urban, elite targets, who are used to hearing call-center voices. However, the youth's impersonated proximity is a subversive (and even possibly insurgent) act that tries to dismantle the unevenness of rural–urban divides.

The digital money from the duped customers' account is routed to e-wallets and the Jan Dhan accounts of the village elderly. The Jan Dhan accounts are part of the Indian government's *Pradhan Mantri Jan Dhan Yojana* (Prime Minister's People's Money Scheme, or PMJDY) to provide bank accounts to the until-now unbanked citizens of the country.[112] The gang of young boys from Jamtara would rent out these new Jan Dhan accounts from the elderly in Jamtara and adjoining villages for a sum of five thousand rupees. Police officials from other states sniffed out the phishing racket when they found money from the credit accounts of Indian citizens living outside Jharkhand going disproportionately into the Jan Dhan accounts of Jamtara's residents. These bank accounts were authenticated through the Aadhaar (unique ID) and connected through mobile phones setting up a "JAM" (Jan Dhan-Aadhaar-Mobile) payment pathway. Given this context, these new mobile phishing scams are constitutive of the "financial inclusion" schemes based on digital money.

The fifth chapter thinks through the interrelationships between financial inclusion and digital inclusion by examining the work of some fintech companies that offer the needy poor in India predatory microloans. These microloans stoke the customers' aspirations but then put them in a debt trap. The phone data permissions initially asked for during the loan onboarding process are similar for both predatory and legitimate loan apps. The predatory loan apps use such data for making threatening calls as part of recovery tactics, while the legitimate lending platforms use it to assess whether a customer with a thin credit file is a worthy borrower based on their social media activity and phone records.

These contemporary lending processes have a longer history in microfinance systems in India and around the world. Ananya Roy and Silvia Federici have critiqued this system, where the world's poor are seen not so much as "risky subjects" who cannot be trusted with loans but, rather, as a new "risk frontier" for potentially being "self-disciplined subjects" who can practice "financial responsibility."[113] Microfinance initiatives have been further transformed by new values attached to algorithmic data and smartphones and what such digital technologies (digital inclusion) can bring to financial inclusion.[114]

Data have become the new collateral for loans given to the economically vulnerable. This sharing of data (including WhatsApp contacts and phone

GPS) for loans (instead of paper documents and a history of credit worthiness) is also seen as a way to evaluate credit worthiness with alternative data footprints, which are not always just financial transactions. With 200 million new smartphone users entering Digital India, new data—including the kinds of contacts one has, the music and music apps one listens to, and the religious guru apps one uses—can be factored into financial data points in the algorithms used by fintech companies such as CreditVidya. Mapping connections between digital lending and digital payments, I situate loan apps within the mobile money distribution ecosystem, including the app stores and the fintech infrastructures operated by identity verifiers and credit scorers. Such is the faith of fintech professionals in the efficacy of data as collateral for lending decisions that a consumer protection researcher ruefully declared to me that, realistically, people have to accept that we live in a "post-privacy world" where "data is part of contract and not so much consensual." She hoped loan apps could just be more candid with the borrowers and no longer ask for user "consent" for releasing data when such parting with data is clearly a "requirement" for receiving a loan.[115]

The concluding chapter, "Limiting: Surveillance, Environment, and Predatory Inclusions," delves into the access vis-à-vis surveillance debate in mobile media studies, where mobile phone users are often asked by the state and corporate actors to submit/release their data and compromise their privacy in order to reap the benefits of recommended entertainment content, access to welfare benefits, and digital cashback and loan opportunities. Data released from billions of Indian mobile phones are stored in data centers to be fetched again later. Data centers are energy guzzlers: extracting value out of data needs energy. Nonrenewable sources of energy are depleting fast, and even "green capitalism" involves grabbing land for solar and wind farms. Environmental thresholds are stretched to meet the logistical demands of mobile media storage and distribution as water meant for agriculture is diverted to data centers' air conditioning operations.[116]

By attending to the discourses around the delivery of government services to its citizens, I track the relationship between mobile media use and political subjectivity. The mobile media user is often made to be the government's ideal citizen subject whose smartphone use releases data for governance and corporate speculation. Many of the fintech and health tech platforms and infrastructures created or facilitated by the state (including the Arogya Setu [health services] app used extensively during the COVID-19 pandemic) use biometrics-based authentication and identification projects that have involved extracting data from the bodies of India's residents and converting them into digital data points mapped/linked across relational databases.

In some parts of the book, even as I invoke mobile phone affordances and aspirational goals alongside caste- and gender-based inequalities in the same sentence, I am not suggesting that the two cancel (or balance) one another. Mobile phones cannot flatten structural inequalities. Rather, mobile phones as technologies are embodied in sociocultural hierarchies and at best can extend social possibilities. The book accounts for the many lives of mobile data in India and their distributions. This is not a feel-good story about mobile distributions but a critical account of the opportunities and challenges, the fault lines and failures in a context in which the regulatory data state and its technocrat supporters have little regard for data privacy and support both surveillance and the expansion of governmentalities through a slew of national mission–level data projects.

1 Loading: Memory Cards

People in Gaya, a city in the Indian state of Bihar speak Magahi, Bengali, Hindi, and Maithili. However, it is Bhojpuri music that dominates music video cultures there. Bhojpuri is a language spoken across the states of Bihar and Uttar Pradesh (particularly Western Bihar and Eastern Uttar Pradesh).[1] Whereas Hindi represents the formal, written language, imposed somewhat homogenously on a large number of communities in the region, Bhojpuri is the colloquial vernacular preferred by local communities.[2]

The resurgence of Bhojpuri music owes much to the proliferation of new television channels that (unlike Hindi newspapers) allow TV news hosts, Bhojpuri film stars, and ordinary people featured on reality TV musical talent contest shows to communicate with each other and the audiences in vernacular Bhojpuri. Beyond television, Bhojpuri *music mania* is sustained by the informal (or, as some would say, "illicit") permeations of such music content through memory cards. When my coresearcher, Abhigyan Singh, and I did our fieldwork regarding circulation of Bhojpuri music in the streets of Gaya in March of 2016, we were greeted by the mellifluous voices of Manoj Tiwari, Kalpana Patowary, and Ravi Kisan and by boards outside shops promoting music downloads on memory cards, colloquially also referred to as "chips."[3]

Later in 2016, streaming began to emerge as a key way to consume and distribute music content in India, but from 2010 until we were walking through the streets of Gaya in March 2016, memory cards were an ascendant form of mobile media consumption and distribution. This chapter on "chip loading"—that is, uploading content to or downloading content from memory cards—is an important account of mobile media distribution history in India.

In the main market of Guraru, a town in peri-urban Gaya, nestled between an electronics store and a men's barbershop ("hair cutting salon" to be precise) is Kushi Gana Downloading (Happy Song Downloading) (see figure 1.1). On a whiteboard, written in neat Hindi handwriting with a brown sketch pen, it is advertised that the shop offers mp3/mp4 versions of music videos, songs, and films with various genres of Bhojpuri music (Nirgun, Birha). Different charges apply for different quantities of songs, varying from 16 gigabytes (GB) for Rs 330 (330 rupees; about $4) to 2 GB for Rs 170 (about $2). Music enthusiasts bring their own memory cards ("chips"), and the songs are downloaded from the shop owner's computer/laptop. These memory cards can then be plugged into mobile phones, and the mobile phone will become the music player, accessing the songs stored in the memory card.

In another location in Gaya, we saw a download vendor without even a semblance of a shop building; he just had a laptop with an internet connection on a table, from which he offered downloads of Bhojpuri songs to people who approached him with their memory cards and mobile phones. This download vendor set up his flexible shop next to hawkers selling *jalebis* (deep-fried batter soaked in sugar syrup) and *kachauris* (a saucer-shaped, fried, spicy snack). The sugary smells of the jalebis, the sounds of the jalebi seller squeezing the muslin cloth to release the jalebi batter in concentric circles on the aluminum frying pan, all mingled with the sound of the download vendor haggling with his customers about a fair price for two hundred new songs. The songs were neatly assembled in a desktop folder on his laptop, titled "Birha" (category of Bhojpuri songs) (see figure 1.2).

Memory cards index the informal proliferations of audiovisual content into subaltern populations. Indeed, even a preliminary ethnography of the Indian informal music economy would indicate that songs and music videos on memory cards (up to 2016) could be found in tobacco and *paan* (betel nut) shops, fast food vans, strip malls, petrol pumps, and mobile charging stations.[4] Memory cards were not just a phenomenon in Indian towns and villages. Mobile recharge shops for cheap feature phones in urban centers were a place where memory cards were sold and loaded with music. Swati Janu has noted how such recharge shops (along with cybercafes and underground gaming parlors) have acted as an "informal digital substructure" for settlement dwellers in cities like Delhi and Mumbai.[5] Rashmi M. has researched how urban subaltern populations, often people who migrated to

Figure 1.1
Kushi Gana Downloading in Guraru. Photo taken by author.

Figure 1.2
Download vendor next to jalebi seller in Gaya bazar. Photo taken by author.

urban centers from villages in search of work like security guards and auto-rickshaw drivers, depended on the "offline media economy" of memory card sharing in mobile phone shops along the streets to meet their entertainment requirements. This act of loading has sometimes been referred to "side-loading."[6] The term "side-loading" could gesture to the fact that the download vendors sit on the side of streets and that they carry out loading of music videos through memory cards as a side-business—that is, side-by-side with other businesses such as mobile repair and stationery goods.

Rashmi M. notes that this informal memory card sharing media economy is essentially a "street economy," something I observed in locations of Gaya and Mewat as well. Digital content in the form of audiovisual treats mingle with edible treats (jalebis and *kachauris*), and because they are accessed by downloads onto memory cards, rather than streaming, digital life here is offline. Digital life is in the streets, part of bazaar atmospherics, and very much an unremarkable aspect of daily life in contemporary India.

Borrowing from Ramon Lobato's work, this chapter discusses the politics of media access and distribution through the heuristics of culturally located informal economies, thereby avoiding the excessive focus on copyright claims and the concomitant "moral drama of property."[7] By attending to the various practices of using memory cards, this chapter tracks the diverse circuits and bazaar atmospherics through which songs and music videos circulate in India. I argue for a research methodology that focuses on the tensions between representation and distribution. In this chapter, I map and trace the informal networks that sustain the distribution of Mewati music videos; at another level I am interested in subaltern-popular representations of Mewati youth in these music videos. More importantly, in these music videos, actors and dancers invoke memory cards and thereby anticipate the future distribution of such videos through memory cards. Concerns of distribution and matters of representation are thus empirically related.

Alongside spatial mobilities of music distribution heralded by memory cards, I shall also be studying the aspirational mobilities of music lovers who consume vernacular music videos. This is consistent with my approach to thinking of "mobilities (un)limited" as entwining aspirational politics and mobile media distribution. Rather than focus on any deterministic effects of media technologies such as memory cards and mobile phones on human beings, this chapter approaches memory cards from the perspective of mobile digital practices. How do people use memory cards? And how do such interactions between music enthusiasts and memory cards lead to new mobilities for digital sound files? The emphasis on *media* practice provides a safeguard from falling into the trap of technological determinism and cultural determinism. Extending the scholarship of actor-network theory, feminist relational ontologies, and subaltern studies, I will be tracing the socio-technical agencies of actors and stakeholders who are part of the mobility network of memory card and music video circulation within the socioeconomic contexts of (Mewati and Bhojpuri) vernacular cultures.

Memory Cards as Part of Mobile Digital Practices

During the period 2010–2016, memory cards had a considerable uptake in usage in India. Memory cards were key to moving the informal (and piratical) practices of music and video consumption from VCD/DVDs to mobile phones. Some music video consumers continued to prefer DVDs on laptop screens over memory cards on mobile phone screens. That said, memory cards made it possible to play a large number of downloaded digital music files on mobile phones. With time, DVD players and TV sets also had memory card slots, but it was the connection between mobile phones and memory cards that mostly propelled the distribution of music videos and songs. The widespread use of memory cards shares a reciprocal relationship with the rise of mobile phones as a popular playback device. There were constraints to playing songs and videos on the feature phones because only a certain number of songs could be stored in the feature phone memory. The memory card or the microSD card solved that problem.

During the time of my research (2010–2016), while mapping digital memory in relation to music in the purportedly "First World" makes us think of platforms like Napster, Tidal, and Spotify, and then about cloud computing and server farms, the story of memory cards in the rural hinterlands of India follows a different trajectory of informal circulation. Digitalization has led to the "micromaterialization" of music; the mp3 phenomenon plays out in different ways in different parts of the world: on the one hand, streaming and cloud computing and, on the other, the flash memory of microSD cards and USB drives (see figure 1.3).[8] Thus, memory cards are part of mobile digital practices, where the "digital" is not synonymous with "internet" or "online" but, rather, with offline media practices.

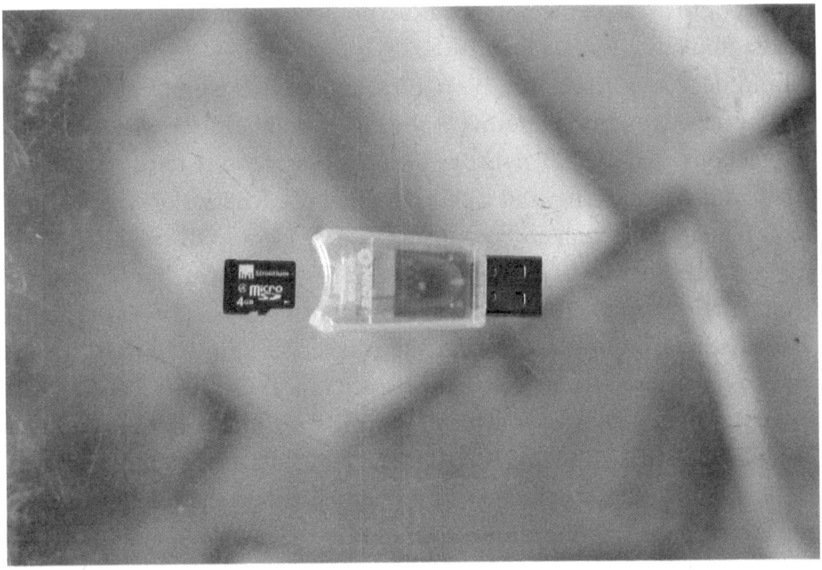

Figure 1.3
Memory card or microSD card. Photo taken by Abhigyan Singh.

Memory cards span different emerging platform systems of audio/DVD players, mobile phones, and TV sets and become part of these various sensory infrastructures. Memory cards reconfigure various mobile media assemblages and create an ecosystem of platforms around them. Little work has been done on memory cards because they are considered part of piratical economies. While some might dismiss memory cards as a transient phenomenon that disappeared with the coming of "seamless" internet connection for all, I argue that memory cards transformed music consumption decidedly from various other forms of audio-video players and audio storage objects to the mobile phone in India. Many research interviewees in Neha Kumar and Tapan Parikh's ethnography of music practices in Bikaner, a town in the province of Rajasthan, India, emphatically declared that CDs and DVDs have been gathering dust since the emergence of mobile phones equipped with memory card slots.[9] In later chapters we will see that the streaming industry idealizes the individualized viewer who can privately watch and listen to streamed content on their phone. One could argue that it was not streaming, but memory card–enabled phone content consumption, that inaugurated the individualized audience.

Part of the reason memory cards do not lend themselves well to digital media studies (or mobile media studies) is because they are part of the digital offline.[10] As a result, the study of memory cards cannot be contained within frameworks of platforms or the logics of algorithms. The socialities, interactions, and patterned uses of memory cards require research methods where one has to go beyond tried and tested applications of computerized tools for data collection and processing. As is evident in this chapter, studying the mobilities involved in the mobile digital practices of memory cards requires attention to offline socio-technical practices. There does exist some related scholarship from Cuba on the circulation/downloading of weekly entertainment packets on USB drives, which is a similar flash memory technology as the memory card, which I will draw upon later in this chapter to suggest the need for thinking about alternative media distribution histories.[11]

This chapter deals with two vernacular music video cultures in India: in the Mewat region, southwest of Delhi, and in the town of Gaya in the state of Bihar, where Bhojpuri music is dominant. "Vernacular" in the Indian linguistic land-time-scape denotes not so much the fault line between the English-speaking and, say, the Hindi- or Assamese-speaking populace, but more so the widening cracks between Hindi-speaking communities and those speaking dialects like Bhojpuri, Mewati, and Maithili.[12] Vernacular music in Mewati and Bhojpuri gestures toward this fault line and asserts particular vernacular identities. The Mewati and Bhojpuri music mania during the period from 2010 to 2016 was sustained by the (micro)infrastructure of the microSD card. In this chapter, I am particularly interested in tracing the shift in the ways that songs and music videos in digital files were accessed, exchanged, and distributed during this period. This marks the beginning of a feeling among rural Indians, even among people who could not afford a smartphone or internet connection, that they could listen to a lot of songs. Songs were like vegetables that could be bought in significant amounts from a market (bazar), and like vegetables, they were essential to life.[13] What circuits of distribution made such a feeling possible? What informal economies of everyday commerce made *unlimited* music a reality?

As we will see in this chapter, memory card–circulated music helped Mewati youth in circumventing the censorship of local video consumption enforced by regional power structures. Mewati youth made and watched music videos that made them aspire to be like the urban youth in nearby cities such as Delhi. Such aspirational mobilities are also unleashed by consumption of "unlimited" Bhojpuri songs and music videos through memory cards in Gaya, and yet caste remains a rigid marker that continues its hold on media practices as social practices, thereby limiting social mobilities. This is the paradox of limits and limitlessness encapsulated in mobilities (un)limited.

Memory cards are part of low-cost mobile media assemblages that, following Ravi Sundaram, could be termed "post-post-colonial sensory infrastructures."[14] Such infrastructures have provided India's subaltern populations opportunities to create horizontal networks that bypass state and corporate power. Let me begin with the download vendor, who is a key node in this network that supports the circulation of music. Much like content platforms today, the download vendor serves as a key intermediary between multisided markets/users.

Download Vendor: Intermediaries and Mediators

The mobile phone shopkeeper or repairman, also known as a *mobile wala*, downloaded music from YouTube or ripped DVDs to extract digital files. The mobile wala then performed the pivotal format conversion on these digital music files. This involved using video converter software to transform MPEG files into mp3, mp4, or 3gp versions. These conversions allowed the music files to be played on mobile phones through microSD cards. The mobile shop attendant in Mewat clarified that the resolution in video files copied from a DVD is high (and, of necessity, these files are extremely large) and is not compatible with being displayed on mobile devices. The shopkeepers would convert the files into various formats, depending on the customer's type of mobile handset. They knew that "x264.mp4" format indicates better video and audio quality, which are compatible with the high-end mobile devices that were then becoming popular in Mewat, while ".3gp" was compatible with older mobile phones.

Earlier, the download vendor used to be a mobile wala, an electronics store employee, or a VCD/DVD shop keeper. Later, the download vendor could just be a person with a laptop sitting in front of a cinema hall. When design ethnographer Abhigyan Singh and I embarked on our fieldwork for the memory card project in January 2016, we saw one such vendor in Gaya sitting opposite the Prem Talkies cinema (in the Dulhingunj neighborhood), where the Bhojpuri film *Jab pyar kiya toh darna kya?* (When You Are in Love Then Why Be Afraid?) was playing (see figures 1.4 and 1.5).

Figures 1.4 and 1.5
Download vendor opposite Prem Talkies working on song folders in his laptop (Gaya). Photo taken by author.

The download vendor had chosen a prime location to set up his shop. Young people from nearby villages who would come to watch Bhojpuri films at the theater would also drop by his store and have their chip filled with more songs, music videos, and films. All that the download vendor required was format conversion software and an archive of songs, videos, films, and wallpaper stored on the hard drive of his laptop. Above his desk hung memory cards, headphones, and audio players for sale.

Figures 1.4 and 1.5
(continued)

The download vendor across from Prem Talkies had folders organized based on musical genre, file type, file size, and date of the songs. He had Bollywood songs and Bhojpuri songs. The subcategories of Bhojpuri songs were *nirgun*, *kajri*, and *biraha*: the differences between them are based on the seasons and moods invoked. The difference, in some instances, is also based on the type of instruments that accompany the singing. The memory card users I saw asked the download vendor to copy Bhojpuri songs, Bhojpuri cinema, Bollywood songs, Bollywood films, and even some photos and wallpaper.

In thinking about the genealogy of the download vendor figure, it is important to historically go beyond mobile repair shops and electronic shops to cassette store owners in the 1990s and early 2000s who would also perform important format conversions and were crucial for music distribution in India. In the nineties, as Jayson Beaster-Jones writes, music from vinyl records was converted into cassettes, and this conversion also involved moving to a comparatively less expensive and more portable medium.[15] Furthermore, the practice of music enthusiasts going to download vendors to have new songs and music videos uploaded to their memory cards also is not some disruptive innovation but part of a longer trajectory of mobile music listening history.

Cassette shop owners realized early on that often only a few (two to three) out of all (ten to twelve) the songs on the cassette of a Bollywood film release would be hits and attract audiences. Hence, there emerged a practice where music enthusiasts would come to cassette store owners asking to have one specific hit (or popular) song added to their cassette. The cassette shop owner would use the dubbing equipment and cue the cassette to the song requested by the customer. There would be a compilation of hits, personally curated for a specific individual, which would change over time with visits to the cassette store. Beaster-Jones writes about such cassette store owners, some of whom were able to transfer their rich archive/stockpile of cassettes onto mp3 CDs, while some others just had to shutter their business.[16] While I have pointed out the similarities between cassette store owners and download vendors in terms of format conversions and meeting customer demands for particular songs, it is important to note that the number of songs on an analog cassette tape would not satisfy the current desire for 2 GB (250) songs but would be restricted to one song or one album.

Facebook, YouTube, and Uber are framed as platforms because they are intermediaries between two different groups of users: between users and advertisers (Facebook and YouTube), and between people needing car rides and those ready to offer them (Uber).[17] In the platform ecosystems created by memory cards, the download vendor can be considered an intermediary between music lovers and music producers. In a way similar to how YouTube connects amateurs and professionals, users and advertisers, volunteers and employees, the download vendor connects song enthusiasts and song producers, and memory card seekers and memory card manufacturers.[18]

In drawing these intermediary-related analogies, there is a collapsing of distinctions between humans and software. Many interactions and connections on social media platforms are generated between different kinds of users and arise because of the work that algorithms perform. However, these algorithms require user metadata (and user data) to keep running efficiently. As Ananbelle Gawer points out, different users of platforms conceptualized as two-sided networks do not have the same rigid functions.[19] Facebook or Google users are not just content consumers; they are also content creators. Similarly, the music enthusiast coming to a download vendor relies on the expertise of the vendor. That said, such a customer might also suggest that the vendor listen to a new song that the customer (as not only a music consumer but also a music enthusiast) has stored on their memory card. If the vendor likes that song, he might upload it from the customer's memory card into his inventory—that is, his laptop's hard drive.

Download vendors are also gatekeepers or curators of content for this memory card platform ecosystem. For instance, we observed a download vendor refusing to load "sexy" music videos for teenagers. Other vendors firmly stated that they do not "download" and offer "adult" content to teenage boys. In such cases, the vendor viewed his work as incorporating ethical concerns about the circulation of certain musical recordings.

Furthermore, platform studies scholars have discerningly noted that Facebook and Spotify like to use the "platform" terminology because even as they devise methods to effectively perform content moderation, they do not want to be held responsible for the content users are making and sharing on the platform. This way, the platform companies also distinguish themselves from media production and distribution companies, which also helps them jump a few regulatory norms that apply to media producers and not platform corporations. Thus, platforms, like download vendors, may not be mere neutral intermediaries but crucial mediators.

The "intermediary" terminology also appears in actor-network theory (ANT) and within ANT in Michel Callon and Bruno Latour's formulation, the "intermediary" as an actor in a network has a less active role than "mediator." If the circulatory network is considered with regard to Latour's ANT, the mobile shop (or the mobile wala) is a key node, a mediator and not just an intermediary. Latour discerns that an intermediary "transports meaning or force without transformation," while mediators "transform, translate, distort, and modify the meaning or the elements they are supposed to carry."[20] Mediators help expand the network into new areas in uncertain ways, while faithful intermediaries aid in sustaining existing networks. The video market was earlier restricted to CDs and DVDs sold in video shops, but when the mobile phones displaced CDs as the medium of consumption, the mobile shop and the mobile wala emerged as mediators. This led to the expansion of the video network. According to Latour, "translation" induces "two mediators into co-existence."[21] The mobile wala became a mediator not by himself but in relation to the affordances provided by mobile phones. Transformations in a network are enacted by the "translation" process set out by the association or concatenation of a set of mediators, in this case, mobile phones and mobile walas. The mobile wala as the human actor is not the sole mediator; the act of mediation involves a complex "intra-action" among mobile walas and various video converters, CDs, and microSD (mSD) cards, entailing myriad discursive-material entanglements.

As a sociotechnical network gets reconfigured, the components of the network enter into new relations with each other, and their functions transform. Karen Barad terms this process "iterative intra-activity," whereby each reconfiguration is a reiteration of the earlier configuration, with slight shifts in arrangements leading to dynamic intra-actions.[22] Abhigyan Singh and I observed similar reconfigurations in the Mewati-video circulation network, when gauging the popularity of mSD cards, the local electronic equipment shops in Mewat started selling new Chinese-made DVD players that play videos from both DVDs and mSD cards. This led to new trajectories of mobility for the memory card data, where if local gossip is to be believed, Mewati videos are being watched by young married couples after midnight on DVD players, when everybody else in the house is asleep.

Espousing a Baradian framework (rather than a Latourian framing), I want to keep this research open to new empirical findings that complicate perceived gendered agencies and that are able to describe how women in the Meo community interact with these videos, how they use their phones and memory cards, and in what ways they interpret these videos. Women in the Mewat district may not frequent the masculinist bazaar spaces of mobile shops, but that does not imply that they do not watch music videos on their phones or with their partners on DVD players. Neha Kumar's work on mobile phone usage by women in India suggests that the gender technology divide might be a matter of perception rather than reality.[23]

I have elaborated in some detail about ANT and feminist relational ontologies because there is no media a priori aspect of memory cards. While the memory cards may constrain or enable particular patterns of activities of music enthusiasts, it is the media practices of human users that shape how memory cards are taken up in the Indian society. The power of memory cards is the power of networks within which there are intra-actions across actants, actors, and agencies constituted by mobile walas, formatting technologies, local entrepreneurs, and music fans.

Before I discuss in detail the memory card phenomenon in Mewat and Gaya, I shall tell a story from the recent history of digital media distribution in the United States. This gives an idea of what platforms would look like if downloading, rather than streaming, were the norm.[24]

Blockbuster and Memory Cards: A Very Short History

Around 2008–2009, Blockbuster's movie rental business was struggling to compete with the Netflix strategy of mailing DVDs to its customers and asking them to mail them back to the company. People were no longer frequenting Blockbuster stores and vending machines where they could rent a DVD. While Netflix had already introduced streaming services in January 2007, those had not been picked up by audiences in a big way yet due to limited internet speeds, and at that time Netflix still advertised its services mainly as "Unlimited movie rental"/"Unlimited DVD: fast." During this interim period, Netflix and Blockbuster assessed and speculated regarding a variety of digital distribution strategies. James Keyes, the chief executive of Blockbuster put forward his hypothesis: "Movies are increasingly going to be downloaded rather than purchased on discs."[25]

Keyes envisioned Blockbuster having self-service kiosks in the company's stores where people could download digital copies of movies into their memory cards. Burning of movies into DVDs in these stores was not a viable option (or satisfying experience) because it would take customers a long time to do so. In proposing this distribution strategy, Keyes hoped that more people would be watching films on mobile phone screens, and such mobile handsets came with card readers. The memory card format was not yet standardized across various electronic media, and Keyes hoped that the entertainment industry and the electronics industry would make the necessary policy decisions so that studios would agree to having movie files on memory cards, and he envisioned that "televisions and Blu-ray players will all have slots for memory cards"[26]

Blockbuster did experiment with two such kiosks in its stores at the Preston Royal Center in Dallas, but the idea did not go a long way.[27] The rest is history, as Netflix revolutionized the movie distribution industry through navigable online interfaces that curate content for users and offer them recommendations based on collaborative filtering algorithms. Netflix, along with Google and YouTube, became one kind of digital platform. In excavating a future anterior (or an alternative) to this dominant platform story, I have gone back to Keyes and Blockbuster's plan. The plan of downloading movies onto memory cards had a short (and unpopular) lifespan in the United States but seems to have had popular afterlives in so-called developing countries, including India, Mali, and Cuba.

Of course, the Blockbuster idea in its Indian incarnation played out differently: there were no proprietary Blockbuster stores, but rather, such music video downloading to memory cards would happen at myriad places (mobile shops, music shops, download markets, multipurpose hardware stores) in each small town facilitated by "download vendors," or "download operators."[28] Crucially, just like Keyes predicted, in India and many parts of South Asia and Southeast Asia, there were indeed television sets, audio players, and DVD players that appeared with memory card slots.

In 2012, the local electronic equipment shops in Mewat began selling new Chinese-made DVD players capable of playing videos not just from DVDs but also now from microSD cards. A microSD card could be put into a USB adaptor, which in-turn was plugged into the USB slot offered by these new Chinese DVD players. So now, DVD players no longer supported just DVDs, and microSD cards did not remain exclusive to mobile phones (and their users). Further, to make the microSD card compatible with new DVD players, various cheap USB adaptors flooded the local market in the Mewat district. I saw some of these DVD players in New Delhi's Nehru Place market, from where they reached Mewat (figure 1.6). Similarly, during a field visit to Gaya in 2016, many shopkeepers remarked on the considerable decline in the sale of DVDs and CDs as a storage format, but DVD players were still being sold, even though not as much as before.

Similar uses of the Chinese portable DVD players can be seen in North Korea, where they are called *notel* or *notetel*, which literally means "notebook plus television." Notel devices have USB and SD card ports along with provisions to play DVDs, and this multifunctional versatility allows notel users in Pyongyang to elude state censorship and surveillance while watching K-pop videos imported from South Korea.[29] Thus, memory cards act here as circumvention technologies. The notel users are being surveilled by the North Korean government, but the government is only able to trace what is being played on the DVD player (often North Korean documentaries) and thus are unable to track what is being played through the memory card (often K-pop).

Figure 1.6
Memory card socket containing Chinese DVD player (Nehru Place market, Delhi). Photo taken by author.

Circumvention Technologies

In local instances in India, similar to North Korea, memory cards are part of circumvention media technologies. According to social norms in Nuh (a town) and Salaheri (a village), if middle-aged and elderly Mewat men are spotted frequenting the public spaces of cinema halls, their social respectability will suffer. Amid such strictures, the memory card–mobile phone combination offers easy and discreet access to Mewati videos. The most prevalent way to circulate these videos in the villages is through microSD (memory) cards, which are procured from street mobile phone vendors. The inconspicuous size of mobile phone screens and memory cards ensures portability and helps youths to privately consume (and spread) the videos. These cards reduce the risk of getting caught by village elders and/or religious men. Rather than keep microSD cards plugged into their phones, some research participants prefer wrapping their cards in small plastic bags and keeping them in their shirt pockets.

Watching videos on television and using CD players for home entertainment have steadily declined with the rise of mobile phones as noted by one of our interviewees: "Mobiles are becoming more famous presently. Everybody is using phones. Every kid has it. Nowadays, even adults have started buying it" (Jyada famous filhal phone ho raha hai. Har log phone zyada use karte hain. Bacche bacche ke pass phone hain. Ab toh bade phone khareedne lage hain). This interviewee also noted that while, initially, Chinese phones had become popular, now there was a demand for bigger phones with more features by Samsung and Nokia: "Pehle Chinese aaye the. Ab to shauk khatam ho gaya. Ab toh bade bade phone Samsung, Nokia le rahe hain" Initially, Chinese [phones] came. Now that interest has died. Bigger phones are coming. They are buying Samsung and Nokia).

The demand for mobiles with expanded screens suggests, among other aspirations, a desire for screen spaces that can better accommodate collective viewing. Watching videos on cell phones is not always a private affair. Young men can get together, away from the eyes of prying family members and relatives, insert a memory card into the mobile phone, and entertain themselves. Even mobile phones they co-own or co-use with other family members can be deployed for this purpose, because the videos stay on the memory card. Moreover, once the card is removed, there is no remaining trace of the viewing activity. One is thus freed from the labor of adding or deleting files in the video library of the mobile phone. Reluctant to visit cinema halls, middle-aged men now carry memory cards.

Personal Archiving Practices

Most consumers of memory cards are neither discerning about nor care whether they are using a legal memory card or a pirated one. Memory cards come in colorful packets with labels that outline the number of photos, time length of videos, and size of music files that will fit on them. Despite the ease and seamlessness of streaming music, the content remains with the content providers (or servers at the data center), which is not the case with memory cards. So, at one level, compared to streaming, there is a level of autonomy enjoyed by memory card holders, and yet some music listeners also complained that download vendors would at times just load what they had. Customers on occasions felt too dependent on a particular download vendor in places where options were limited.[30]

People have a personal attachment to their microSD cards; they keep them safe from scratches, they love to talk about their content (unless some of it is censored/pornographic), and as such, it perpetuates a personal archiving practice. Our interviews with locals in Gaya suggested that these personal archiving practices are marked by a variation in the frequency of visits to the download vendor (interview excerpt, Gaya, January 2016):

INTERVIEWER: How many chips do you have?

TAXI DRIVER: I have an 8 GB and also a 4 GB [chip]. I use one of these on my mobile phone and the other in [the music player in] my Jeep.

INTERVIEWER: How frequently do you load your microSD card?

TAXI DRIVER: Well, I load my card once a month or two.

CONSTRUCTION WORKER: Once a week.

While some customers were inclined to retain their favorite content during the reloading process, others preferred a completely new set of songs and music videos, and during their next visit would ask the download vendor to completely "format" (that is, delete existing content from) the card. Based on our interviews, we also got a sense of the loading preferences and the different priorities and considerations that guided particular preferences. In some cases, one had to choose between packing in a lot of low-quality mp3 music and low-resolution 3gp video files or a smaller number of high-resolution music videos and high-definition films that took up more space on the memory card. The following excerpt is from another interview (Gaya, January 2016):

INTERVIEWER: How much do you pay for complete loading of an 8-GB card?

TAXI DRIVER: Fifty to sixty rupees to load all songs [USD 0.77–0.92].

INTERVIEWER: And how much for uploading films?

TAXI DRIVER: [Uploading] one film could cost from five to ten rupees [i.e., USD 0.08–0.15]. An HD [high-definition] film costs ten rupees.

Memory cards as storage technologies are shaping vernacular languages and cultural memory. Downloading and uploading songs/videos onto memory cards are referred to as "loading." And, as has been said before, memory cards are colloquially termed "chips." Chip loading therefore is a practice very much tied to mobile listening and personal archiving. Indeed, what I have found in my own research fieldwork with Abhigyan Singh, and what later, Deo and Duggal also noted in their writings, is that memory card loading (uploading/downloading) is about purchasing not one album or song (something so prevalent in the freemium/premium model of iTunes) but, rather, a whole assortment of audiovisual content in an aggregate of gigabytes of data, just as one would buy vegetables from the market, quantified in kilograms.[31] It is important to reemphasize that not only is the literal context of such a digital purchase an open-air market but the mode of purchasing is also similar to buying vegetables. Neha Kumar, in her research about memory cards in Bikaner (Rajasthan) and Maksi (Madhya Pradesh), notes that there was a tendency among music listeners to try to outcompete one another in terms of the number (quantity) of songs and not necessarily the quality of songs stored in their memory cards.[32]

Circulatory Assemblage: Mobility of Digital Media Files Offline

Extending across the states of Haryana and Rajasthan, the Mewat region lies southwest of Delhi. In Mewat, the fieldwork was done in the Nuh district of the state of Haryana and village of Salaheri within that district. In Nuh, listening to radio is popular, and community radio stations such as Radio Mewat have attained considerable popularity. That said, *maulavis* (religious preachers) prescribe a strict objection to the consumption of visual images by restricting newspaper readership, Bollywood film watching, television viewing, and digital editing. Within this visual censorship regime, supposedly notorious Mewati videos had gained considerable popularity around early 2010s. These music videos, in the form of mp3 files, move through a network whose actors and stakeholders I will trace in this section.

The offline mobility of Mewati music video files depends on format conversions performed by mobile walas and on the movement of traveling workers like truck drivers and taxi drivers. The Mewati video-circulation network begins with the producer, who invests money to generate Mewati video CDs. Each CD contains roughly two to four videos. "Dealers" purchase the CDs from the producer to then distribute them to local mobile shops. In a month, a dealer buys about five hundred CDs for distribution. Mobile walas (mobile store owners and attendants) explained that these videos first arrive at peri-urban markets in Nuh and later make their way to villages such as Salaheri.

The mobile walas copy a CD onto their laptops, desktop computers, and hard disks, which given the digital materiality of music files, they only need to do once—paying the dealer anything between 20–30 rupees for allowing the reproduction. Every month, on an average, the mobile walas copy about fifty new CDs. The CDs are often called "cassettes." The mobile shops in Nuh can be found in the local market area, where they offer mobile repair services, upload digital files onto USB sticks, and sell mobile hardware. Along with memory cards, they sell SIM cards, Mewati cinema, Mewati videos, videos of stage shows (music and dance) from Pakistan, and "pirated" Bollywood and Haryanvi songs and movies.

Many local Meo men work as taxi and lorry drivers, and they carry the Mewati video CDs and loaded mSD cards with them. Our research participants explained that because of these lorry (or truck) drivers, Mewati video CDs are available across India in places as far away as Mumbai and Goa. These trucks mostly do not have screens, so the drivers listen to the audio part of Mewati videos. According to our interview participants, some new "taxis" (cars, jeeps) now have a small (LCD) screen that makes it possible to play videos using mSD (through a USB connector). The Meo youth exchange mSD cards and use Bluetooth to share files across mobiles, especially among their friend circles. There is no strict end user in this circulatory network, as every user is a potential transporter or circulator. Given this configuration of the circulatory network of Mewati music videos through memory cards, it is not surprising that truck drivers are one of the primary target audiences of these music videos, and their presence in the music videos is not merely incidental.

Let's look at the album *Subin sher Chandanki ke* (Subin, the lion of Chandanki). Chandanki is a village in the Mewat district, and Subin is a locally popular lyricist. Both the singer and his album are advertised using his name. In the introduction to the music video (or the opening credit sequence), a voice-over (most likely by the singer himself) notes, "Yeh album Mata company Dilli ke driveron ve unke mechanikon ke sahyog se paish ki gayi hai" (This album is presented in cooperation with the Mata company of Delhi's drivers and mechanics). This advertisement (for the Mata company of Delhi) appears in the music video in all likelihood because some of the company's drivers and mechanics come from Mewat.

Subin sher Chandanki ke starts off with an explosive song. The song is accompanied by rapidly edited shots of Asmeena matching steps with the female dancer, Jyoti Alwar, then with the male performer, Salim, and then again with Jyoti, but in a different costume and location (see figure 1.7). Eight minutes into the video, Salim kneels on one knee, holding Asmeena's hand as she dances in a circle, and we hear him singing, "Barah-battis-baavan gaadi duniya main mashoor, company Mata hain pyaari" (12-32-52 vehicle is popular in the whole world, Mata company is lovely). As we hear

Figure 1.7
Salim and Asmeena. Screenshot from *Subin sher Chandanki ke.*

Figure 1.8
Mata truck company advertised in Mewati music video. Screenshot: *Subin sher Chandanki ke.*

these lines, the sequence alternates between low-angle shots of a board with the registration number "12-32-52," of the Mata truck company, and of Salim and Asmeena dancing (figure 1.8). The truck is praised on the audio track—"hawah main poochi lehrari"—suggesting that the vehicle moves like waves in the wind. The next shot cuts from the dancing couple to a truck weaving on the road, thereby enacting the wavelike motion sung in its praise. We hear Madam Chanchal's voice croon, "Fakru aur Sirazzuddin donu masti leeyera, Chandanki ka kama ke diyeda" (Both Fakru and Sirazzuddin [two males] are having fun [and] earning and giving to Chandanki). This portion of the lyrics most likely suggests that Mata truck company is a reliable employer because it provides a satisfying work experience and guarantees good wages to the youths of Chandanki.

One of the characteristic features of these music videos is that there are alternating movements between song-and-dance routines and advertisements for sponsors. The representation logics enfold the distribution logics as the key distributors of Mewati music videos (truck drivers) are addressed (and represented) in these music videos. In a way, during production (including editing), the Mewati music videos already anticipate their distribution through memory cards carried by truck drivers. Transportation modes and communicative media fuse together as representation and circulation entangle themselves, a characteristic we will see forming a pattern with further examples of walking and motorbikes related to memory card devices in the course of this chapter. These interdependencies across and conjoining mobility systems of transportation and communication add richness to the theorization of mobilities (un)limited.

Subaltern-Popular Representations: Limitations of Aspirational Mobilities

Meos living in Mewat are erstwhile Hindus (Gujjars, Meena, and Rajputs) who converted to Islam centuries ago yet continue to preserve their beliefs and practices from the pre-Islamic past. As a cultural community, Meos thus maintain a distinctly liminal identity. They were subjugated throughout medieval and recent history by successive state regimes (Mughals, Rajputs, British, and Indian state) due to a variety of socioreligious issues, including their unique identity, which does not fit into the modern Hindu-Muslim binary divisions.[33]

The Indian state of Haryana has marginalized the Meos as a community; they do not even have a durable water supply.[34] However, they do have access to and use popular-media representations (music videos) to make political and identity claims, challenging hegemonic forces, and publicize them through circulation networks. I read the emergence of vernacular identities alongside the speculative heuristics of the "subaltern-popular" to explore the question, Why express desire and aspiration when they are socially censured?[35] The Mewati boys (*choras*) would like to force a comparison between themselves and their vernacular identity, on one hand, and the elite middle class living in urban locations like Delhi, on the other. Hence, we have the figuration of the Pulsar motorbike or the Bolero jeep, which asks the Mewati audience to aspire and imaginatively travel from Mewat to Delhi and yet invites the reprimand that the bike, or the jeep, most likely was stolen from Delhi.[36]

The analytical category of the subaltern, in the Gramscian vein, considers marginalized classes on their way to acquiring hegemony through collective struggle. However, the hyphenated concatenation of the subaltern and the popular is not without its set of generative tensions. The "Gramscian assumption" that the "subaltern is defined by insufficient access to modes of representation" makes it difficult to imagine how access to popular media might help the subaltern.[37] Furthermore, some subaltern studies scholars suspect whether popular-media representations could help forge collective agency. For Chattopadhyay and Sarkar, it might be useful "to think of the popular as the subaltern on its road to hegemony (and therefore representation)."[38] By expressing aspirations and desires in their own idiom, Mewati videos potentially invite theorizing about the subaltern with the popular as one comprehends video cultures. The subaltern-popular framework enables understanding of how marginalized communities use access to popular-media representations to challenge hegemony. In my analysis, I explore cultural practices that consider "subaltern" in a relational sense, and not in its radical alterity where, as Gayatri Spivak would say, "subaltern is removed from all lines of social mobility."[39]

Publicity is the key driver for the production and circulation of Mewati videos. Willing male participants who want to be featured in the video provide the production money. Participation, framings, and screen time are interconnected: 5,000 rupees for brief screen time (usually between ten and thirty seconds) and much more to dance around specific female dancers. Whether it is being seen as a well-earning man or as flaunting motorbikes and cellphones or as a man surrounded by women, through the music videos, young men of Mewat seem to want to achieve a certain status of masculinity, for masculinity is never an ascribed status but something that has to be earned by repeatedly demonstrating a sense of superiority.

In Mewati videos, Meo youth pose with motorbikes and mobile phones as songs with suggestive lyrics play on the audio track (figure 1.9). Such posing indicates an aspiration to catch up with the neighboring urban youth of New Delhi but also earns the rebuke of village elders for spending family's savings on a motorbike.[40] In December 2012, I was conducting research with Abhigyan Singh in Mewat about the use of memory cards/microSD cards by Meo youth for listening to songs and watching music videos. During that time, I distinctly remember the taxi driver who was taking us from Delhi to Nuh (in Mewat) telling us, "Sirji, why are you going there. It is a bad place." The elders among the Meo community were worried about these music videos. When asked to explain the reason for their anxiety, the Meo elders noted that people might say that the motorbikes these *choras* (lads) pose with in the [Mewati] videos were stolen from Delhi. Our taxi driver, on his own volition, confirmed this when he said, "Sirji, whenever there is a bike stolen in Delhi, the police know they should come and search in Mewat."

Figure 1.9
Posing with motorbikes in music videos. Screenshot: *Subin sher Chandanki ke.*

These Mewati music videos, as this analysis shall bear out are as much about regional identities as they are about "transregional mobilities"; they are as much about the politics of language as they are about the politics of development.[41]

Scholars like Chamberlain and Witteborn define aspirational mobility as the desire and hope to "make it"—that is, to better one's life in material and social terms.[42] The aspirational mobilities of Mewati youth manifested in their desire to live lives like urban Delhites riding motorbikes and using sleek phones is shaped and enacted through mobile phone cameras and memory card–enabled storage and circulation: such mobile digital technologies make possible the "representation, witnessing, and archiving of desires and imaginations."[43]

However, this aspirational mobility of Mewati youth has limitations, as we see that even as they aspire to the urban lifestyle of Delhi, the representations of them with jeeps, bikes, and mobile phones backfire. The geographical mobilities of bikes and mobile phones also end up marking Mewati youth as possible thieves and criminals who do not quite own these sensuous mobile objects but have stolen them; their social mobility remains contained.

It needs to be stressed that the desired practice of posing with motorbikes in music videos is certainly not restricted to Mewati youth or Mewati videos.

Reels on short video platforms have been the new venue for the vernacular creativity of regional music videos, where one continues to find youth posing with motorbikes. *Dupahiya* (2025), a recent OTT show on Amazon Prime scripts the associational aspirational mobilities of motorbikes and mobile phones. In the show, Roshni, a village girl in the state of Bihar is getting married to a guy who has a job in the city. The groom demands a motorbike as dowry. The girl's parents agree: they calculate that the investment in the bike will take their daughter from village to the city. Roshni's brother Bhugol, even though he does not know how to ride a bike, poses with the newly bought bike to make an InstaReel, hoping the video goes viral and he can become an influencer and then eventually a star. However, while shooting the reel, Bhugol ends up losing the bike, as it gets stolen from him. The motorbike, or *dupahiya* ("two-wheeler") for which the show is named ends up becoming the whole village's obsession and drives the show's plot.

Caste (Im)Mobility

Although microSD cards are usually used with mobile phones to listen to music or watch music videos, other possibilities in the form of memory card–enabled iPod look-alikes and can-case audio players have emerged.

New Chinese-manufactured audio players sold in rural Gaya (in India) are made of metal cans or plastic cases from Gillette shaving cream or Coca-Cola equipped to play music from USB sticks and microSD cards (but not CDs or cassettes). Other casings include those from deodorants such as Axe, known as "perfume" players. Some of these casings/cans also have images/scenes on them from the latest Bollywood films, including faces of Bollywood actors (see figure 1.10). What the soft-drink, deodorant, and Bollywood can cases have in common is that they are all audio players that possess the same basic circuitry/foundation (think of a platform as a foundation): identical configurations of slots/ports for microSD/TransFlash cards, USB thumb drives, headphones, chargers, and auxiliary cable ports for connection with mobile phones, tablets, and car stereos. Thus, the basic foundation, the platform, is the same even as different can-case players have different affects and facilitate listening to popular vernacular songs. Again, the Chinese aspect of these DVD players is crucial: the *mobile wala* who helped us see the internal circuitry of these audio players mentioned that even the screwdriver he used to open them was made in China (the screws, too, were Chinese) (see figure 1.11).[44]

Figure 1.10
Can audio players (Ali Bhatt (Bollywood star), Coke, Axe). Photo taken by author.

Figure 1.10
(continued)

Figure 1.10
(continued)

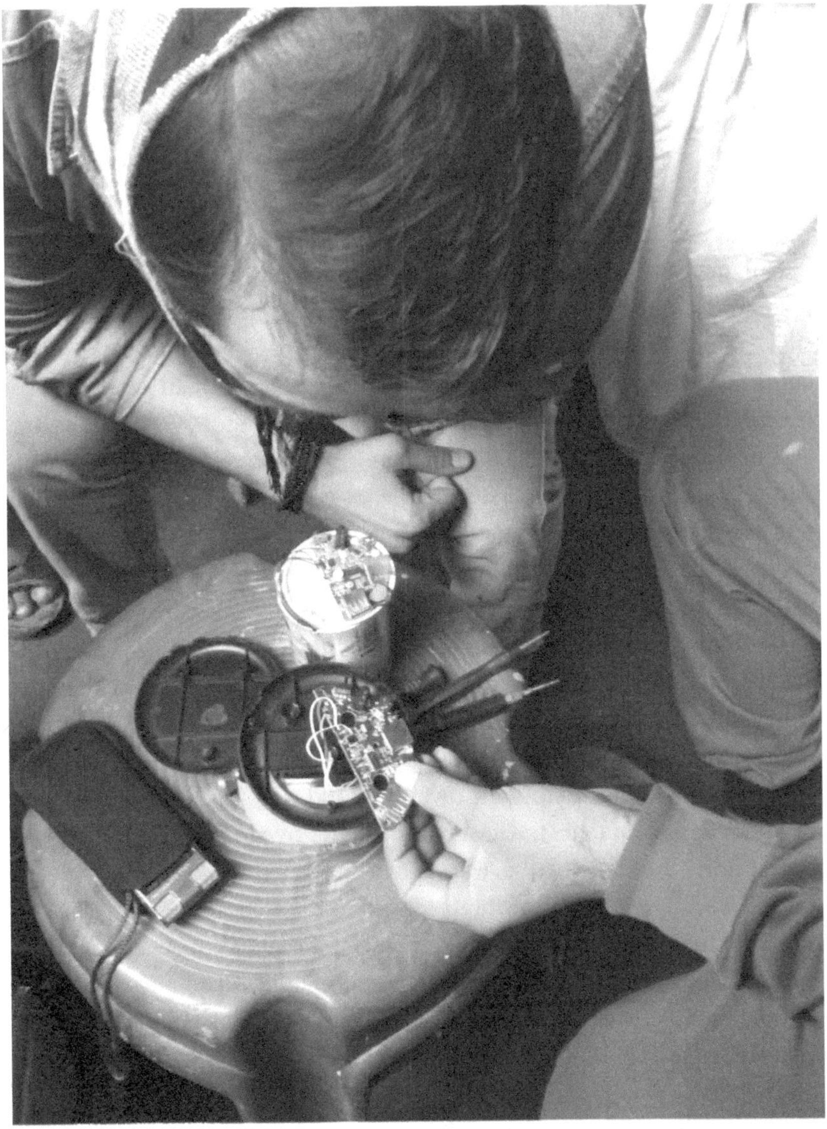

Figure 1.11
Made in China screws used in memory card–enabled audio players. Photo taken by author.

For many users, the mobile phone (into which the "chip" fits) is an object of intense emotional investment. Since one tends to carry the mobile phone everywhere and use it for so many different purposes, mobile phone users think of it as part of themselves, as part of their performance of themselves, and as extensions of their bodies. Moreover, mobile phones can be customized in many ways, by modifying features and affordances such as ringtones, wallpaper, and case.

The memory card–enabled iPod look-alikes and Chinese audio players are portable but do not have as many features/affordances as phones and thus are not as popular. That said, mobile phones have inaugurated taste cultures, fashion trends, and aesthetic evaluations, which seep into the Chinese audio players. A new kind of public mobile music listening culture is being inaugurated when a young man is strutting on the streets of Gaya with a Pepsi can-case audio player. Depending on whether he is carrying an Axe can-case audio player or a Kingfisher (a beer company) can-case audio player, or a *Piku* (Bollywood film) can-case audio player (see figure 1.12), he will be judged in different ways by onlookers, and his persona as the owner of the audio player will be evaluated differently.

Our interviewees, who were often electronics store owners, disclosed that when they first emerged in the city, these audio players were considered trendy and were even used by yoga instructors in their classes in New Delhi. However, interest in them soon dwindled, and they became "kids' things" or toys. The respondents in Gaya said that after five to six months, the only people using the audio players were people of low caste (*neechi jaat*) or "small caste" (*chhoti jaati*) or those who belonged to the "laboring class." The consumption politics of these "cheap," "chip"-enabled audio players in a place such as Bihar cannot be divorced from caste politics. The affective impact of holding a Pepsi or Kingfisher can-case audio player in public should be read within the context of taste cultures, which are heavily marked by caste hierarchies.

Gaya is located in Bihar, which is known for its hierarchical caste system and for people who invoke caste hierarchy to describe taste cultures as part of quotidian conversations. Our high-caste interviewees presumed that the Chinese audio player, which was inexpensive, was used by people in socioeconomically backward castes. Consumers of audio players were seen as those who could not afford mobile phones but required good sound quality to enjoy listening to Bhojpuri songs. This kind of caste-based

Figure 1.12
Discussing the act of mobile listening on portable audio players just outside Plaza Electronic Complex, downtown Gaya. Photo taken by author.

explanation of taste cultures was not provided in Guraru (a neighborhood in Gaya with a considerable rural population), where people from various castes had bought these players. We also observed construction workers, both in Gaya and Guraru, using these audio players so that they and their coworkers could collectively listen as they worked. Here is one sample of responses we received about the profile of users of the Pepsi can-case audio player (interview excerpt, Gaya, 2016):

SHOPKEEPER 1: These [Pepsi players] are bought only by *chhote log* [small people], like Harijans [Dalit]

RESPONDENT: *Chhota jaat* [low caste] like Bhuiya and Chamar use such things. You will see these in small and poor people's homes. They cannot afford TVs, so they keep and use these things.

Another shopkeeper stated, "Mostly 'low-quality' people buy these." When asked to clarify what he meant by "low quality," he replied, "Low-quality people such as labouring class. They keep the player wherever they work. So, they listen to music and also keep doing their work." Another shopkeeper specified, "If *we* [educated people from a "better" socioeconomic class] roam around holding this 'Pepsi' can, it will look absolutely rubbish. Everyone will ask, why are you holding this?" (italics, my emphasis).

Bhojpuri music and cinema are entangled with altercations regarding caste and debates over the vernacular challenge to the hegemony of Hindi in the Indian heartland. Ratnakar Tripathy explains that lower castes in the state of Bihar have gained from their representation in political parties and their sociocultural representation in new Bhojpuri cinema and music.[45] These lower-caste communities have not only culturally asserted themselves but also aspire for a middle-class status and lifestyle. The upper-caste communities (Brahmins, Thakurs) have recently found their political representation in Bihar slipping, and they begrudge the rise of lower castes and despise their aspirations for joining the middle class. Such caste politics are also entwined with the affective platforms that enable listening to Bhojpuri music. The consumption politics of these "cheap," "chip"-enabled audio players in a state such as Bihar cannot be divorced from caste politics. The affective impact of holding a Pepsi or Kingfisher can-case audio player in public should be read within the context of taste cultures, which are heavily marked by caste hierarchies.

Like in the example of the Mewati music videos, I want to draw relations between mobile media, the geographical mobilities they afford, and the socioeconomic mobilities users engaging with them aspire to while discussing the ways in which Bihari Dalit youth use memory card–enabled portable audio players in the form of Pepsi or Axe cans.[46] Tim Cresswell asks us to think of the "politics of mobility" in terms of "material movement, representation and practice."[47] There is a politics to this digital mobile practice—the embodiment of the memory card–enabled portable audio player in the hands of the Bihari Dalit boy, the entanglement of caste and gender as the Dalit boy walks through the streets of Gaya listening to Bhojpuri music on a Pepsi-can audio player. He desires, or still better, aspires to a social mobility from this listening act, which is denied to him in the larger discourse about such mobility by upper-caste men of Gaya. There is an assertion of caste, and the digital listening act is a way of expressing that socioeconomic mobility and taste cultures will no longer be constrained by caste.

The Dalit boy walking through the streets of Gaya holding a Pepsi-can audio player is forging a connection between novel listening practice and emerging social relations (beyond caste), and yet in representations of such mobilities of the Dalit boy by our upper-caste interview participants, caste remains an entrapment, and socioeconomic mobility a myth. So, indeed, like the case of Mewati youth posing with mobile phones and motorbikes, there are limitations to the aspirational mobilities of Bihari Dalit youth as well. That said, aspirational mobilities here need not be contained by the regressive opinions of our upper-caste interlocutors. The memory card–enabled portable Pepsi audio player belongs to a section of society that aspires to make it to the new middle class, thereby challenging the values of higher castes who have historically occupied the middle-class position.

Conclusion

The use of memory cards and download cultures that I have discussed are not happening just in India; they can be also observed in places like Cuba, where every week people have their memory sticks filled with the latest internet "packages" (*paquetes*) containing films, American television series, software, and even ads for local Havana hotels/restaurants.[48] The *paquete* has been a hugely influential media object that has shifted distribution assemblages in a significant way in Cuba. Laura-Zoe Humphreys argues that "*paquete* shifted control over distribution in a large-scale fashion out of the hands of the state for the first time since the onset of the Cuban Revolution in 1959, facilitating widespread circulation of both global entertainment genres and new forms of local cultural production."[49] This popularity of memory sticks or USB drives in Cuba should be understood within a context where "online activities remained limited throughout the 2010s," making it difficult for people to surf the internet unless they found public Wi-Fi in parks under the duress of uncomfortable heat.[50]

Similar to the pivotal role played by the mobile repair shops, download vendors, and taxi drivers/bus drivers in sustaining the memory card–enabled distribution system India, the memory stick (*paquete*) media economy in Cuba is also based on key mediators, intermediaries, and specific social spaces and transportation modes. It was run by *matriz* (original source)/*paqueteros* who would download videos from internet/YouTube or record them via illegal satellite connections that would be saved in multiple hard drives and then transferred to weekly packaged USB sticks with specific labels. After that, they would be delivered by messengers to major clients in Havana and to interprovince bus drivers who would take them to resellers across the island.[51] Geoffrey Hobis has written about how temporary labor migrants use microSD cards to store, consume, and transport films from urban areas to remote villages in the Solomon Islands.[52] Like Hobbis, I have argued that mobility of data is intricately linked to physical movement of human bodies.

Earlier in this chapter, I mentioned some histories of analog-digital connections in terms of audio cassette–related media practices finding new forms in memory card–mobile phone–enabled music listening and curating practices. Debra Spitulnik, among others, has theorized cheap audio-recording media technologies that are portable and small, particularly audiocassettes (including audiocassette recorders) as "small media."[53] These small media are a hybrid form of mass media and interpersonal media, more appropriately thought of as alternative media. Such media are deployed by small groups to express themselves, not necessarily for political mobilization but to foster group solidarity and identity formation in many parts of the world, from Iran to Zambia and from Nigeria to India. The memory card and mobile phone combination (even as they are digital media), like analog small media, are portable. This chapter also suggests that many people use mobile phones not just to communicate but also as a digital assistant, which can play sound and video files.[54]

O'Neill et al. argue that although in 2009, microSD cards and Bluetooth were key file-sharing technologies, by 2015, because of the easy availability of low-cost smartphones, WhatsApp has become the primary sharing app among even lower-middle-class groups in Indian cities such as Bangalore.[55] Much of the study that I conducted with Abhigyan Singh on memory cards in Mewat was in 2012, and since then much has changed, including among many of our interview participants, who now use WhatsApp to chat with us. An app version of the memory card has been developed by a Chinese platform company SHAREit, which basically app-ified the memory card functions and catered to rural India, where people started sharing videos through the app rather than memory cards. Realizing that some users in rural India go online less frequently, say once a week, SHAREit changed its algorithm to work on a delayed feedback mechanism for such customers.[56]

That said, during our January 2016 fieldwork in the peri-urban regions of Gaya, we still observed wide use of memory cards on feature phones for music sharing/listening among the working class. Furthermore, as discussed in this chapter, memory cards are useful circumvention technologies in North Korea and India because their use does not leave any digital trace. Also, someone who has one 8-GB memory card with 20,000 songs owns a personal music archive. The archival culture around memory cards would suggest that it involves a different kind of music-sharing experience than using Bluetooth or WhatsApp. That said, Bluetooth, WhatsApp, and memory cards are all part of intersecting sharing economies and participation cultures, some of which we will discuss in other chapters.

Across the various chapters of this book, I discuss the rise of aspirational mobilities in India. In this chapter, I have elaborated on the freedom and constraints of particular aspirational mobilities for Mewati youth and for lower-caste music enthusiasts in Bihar who use memory card–enabled mobile/portable media. These mobilities have much to do with both communicative media and transportation systems, a phenomenon encapsulated in the concept of mobilities (un)limited.

After this chapter about informal offline data-sharing networks, we move to formal online streaming networks in the next chapter, which deals with how post-2016, streaming media industry professionals targeted the new mobile internet consumers/users in a very focused manner in order to ensure that they do not watch pirated/download videos. The next chapter therefore examines a very different kind of distribution system with very different goals and a different relationship to piracy, copyright, and informality. Despite that, I argue they also, like the memory card–enabled offline distribution assemblage, involve human and nonhuman mediators/intermediaries engaged in complex socio-technical practices, which offers further opportunities to conceptualize mobilities (un)limited.

2 Streaming: (Un)bundling and Deep-Edge Computing

The rapid increase in video-on-demand (VoD) streaming services brought forth significant shifts in processes of production, consumption, and distribution in India. As India has moved toward producing and consuming more and more streaming audio-visual content, there has been a concurrent rise in internet-enabled mobile phones (and, in several cases, smartphones). During my fieldwork in June 2017 in the city of Mumbai, media industry professionals had already begun thinking of smartphone users who could stream one to two gigabytes (GB) of data per day as their possible target customers. This feeling only firmed up through the years of intense COVID-19 lockdown, when the demand for streaming content swelled. If the new mobile phone users (neomobiles) were a key audience that the VoD providers were trying to reach, what was their distribution strategy going to be?

By the time I met some of these industry people again in August 2022, streaming platforms like Hotstar and SonyLIV had already worked out deals with telecom companies such as Reliance Jio and Airtel to have their services be part of bundled offers. Airtel and Jio had emerged as over-the-top (OTT) aggregators, meaning they offer media directly via the internet, bypassing traditional cable services and aggregating streaming apps together. The Airtel Xstream feature aggregated various video apps with a single log-in, subscription, and search. Variously valued packs of streaming-service apps were available through both Jio and Airtel. One could subscribe to all the different streamers—Disney+Hotstar, Netflix, ALTBalaji, and SonyLIV—or sometimes just one or two in a less expensive pack. It should not be surprising to recognize that telecom bundling as a strategy is popular not just in India but in many countries of the Global South, including South Africa, where the local telecom Vodacom offers Prime Video and regional streamer Showmax as part of its mobile packages.[1] Telecom operators are key business partners of streaming platforms in several countries of the Global South because mobile phone subscribers are potential customers of video platforms, and the kind of mobile phone and mobile internet connection shapes several streaming considerations of reception and payment.[2]

Around 2015, both streaming content creators and streaming distributors (especially the telcos) were gearing up for a streaming ecosystem in India. Industry insiders have described 2015 as the time when the winds of change were being felt. More and more creators moved from television-network show production to making web series on the internet, and streaming content became, slowly but surely, a thing to produce and distribute.[3] Hotstar was officially launched by Star India on February 11, 2015, coinciding with the 2015 Cricket World Cup and upcoming 2015 Indian Premier League (IPL). People no longer had to sit at home watching cricket on television but watch view it on their phone while they were in a car or train, on their way home from work. Soon after 2015, the digital television landscape changed significantly with the entry and dominance of streamers like Hotstar and Amazon Prime Video, followed by Netflix and other Desi/Indian streamers like SonyLIV, ZEE5, and ALTBalaji.[4] From just one streamer in early 2015, the number of national and regional video-streaming platforms increased to about thirty in 2017. By early 2022, there were about fifty streaming platforms catering to domestic and international (including diasporic Indian) audiences.[5]

The distribution landscape was also shifting with the entry of Reliance Jio as a new cellular operator in 2016. The company is credited with an infrastructural overhaul of the existing cellular industry by emphasizing 3G and 4G long-term evolution (LTE) data-services technologies. To remain competitive with the new entrant Jio, some of the incumbent telecom players like Airtel, Idea, and Vodafone also invested in improving cellular infrastructure, especially 4G LTE technology.[6] These shifts by the telcos helped realize the streaming ecosystem from the distribution side.

From bit rate ladder optimization to deep-edge computing, this chapter discusses various techniques used by content curators, encoders, and content delivery networks to make sure that the audience that has recently acquired a mobile phone ("neomobile") is able to have a seamless streaming experience with no buffering. To have mobile users experience the convenience of low latency times is a promise or desired objective of corporations, an instance of what I call "mobilities (un)limited." In the context of logistical considerations about mobile media distributions, mobilities (un)limited has a lot to do with both instantaneous transmissions and geographical movements. Repeatedly in my conversations with content distributors, they focused on a scenario where the challenge was to provide seamless service to a mobile phone user while they were moving—in a train or auto-rickshaw during their work commute. The discussion of mobilities here, then, is about "interdependent mobility systems, infrastructures, and mobile capacities," including mobile phones, transportation systems and trains, automobiles, and automobilities that reconfigure fragmented lived experiences.[7] Finally, the talk of limitless mobilities has to meet the challenges of relatively sedentary (or immobile) infrastructures, like the data servers where all mobile internet data goes. Media infrastructures like these work with energy sources that are scarce or difficult to procure, so there are limits to the mobile streaming story as well. As we will discuss, there is this narrative in the streaming industry to try to bring relevant content closer to the user, also referred to as a movement away from the center (data center/cloud computing) to the edge (edge computing/fog computing). Some local content delivery networks (CDNs) espoused a deep-edge computing (hyper-local) model and worked with small, local internet service providers (ISPs) (instead of big telecom players). They stressed the importance of understanding local neighborhoods and their language preferences and taste cultures in order to effectively distribute in a more targeted and optimized manner. These small local ISPs in most cases were run by a resident of that specific neighborhood, with an organic link and ground connection with the community, thereby harking back to the cable-walla (cable operator) figure of the erstwhile cable television distribution model.[8]

It is crucial to understand the shift from a mobile media distribution system predicated on portable memory cards (chips) procured from mobile repair shops or download vendors (mobile-wallas as discussed in the previous chapter) and inserted into mobile phones to that of streaming data from data centers through content delivery networks to mobile phones. While streaming presumably makes a wider set of media content available to those mobile phone users who can subscribe to or pay for it, the control over that content is much less, even as streaming content is often understood as "on-demand." The autonomy of the mobile phone user is much more at times in the case of memory card distribution logistics. That said, as I describe the operations of local CDNs, streaming distribution no longer always seems to require the content to come from a data center far away but might be closer to the user, stored in relatively smaller cached storage locations, which could be a 5G cell antenna site, neighborhood café, or school building. There remain comparative questions of the convenience and inconvenience of streaming and downloading, and there remain discussions of the formality and informality of distribution industries as we move from a chapter on memory cards to one on streaming infrastructures.

Internet video services have a different distribution system than a direct-to-home broadcast system through set-up boxes and satellite dish antennas. The video format supported by satellite signals does not go through an internet pipeline. This entails encoding a raw file to a digital format and packaging it across multiple bit rates. I was told about this by streaming platform executives who had earlier worked with legacy TV companies and then had to start launching subscription video-on-demand (SVoD) and advertising-based video-on-demand (AVoD) platforms in 2015.

The central question for this chapter is this: How do streamers and telcos envisage their new mobile phone audiences, and how are they tailoring their streaming infrastructures to target such audiences? Other questions emerge about the kinds of challenges faced by TV executives in transitioning to serve streaming content to subscribers on their mobile phones or about the nature of the dependence of streaming platforms on content delivery networks. Audience imaginaries shape and are shaped by streaming distribution infrastructures, and the center/edge dynamic prevalent in the discourse of content delivery networks suggests the importance of the local, whether it is local language and taste cultures or local data storage capacities.

In a related vein, other questions arise: How do streaming distributors take into account India's varied bandwidth geographies (across tier 1, tier 2, and tier 3 cities/towns) and linguistic/regional diversities as challenges to their distribution operations? How do they imagine their mobile televisual viewers (in all their linguistic, gendered, caste, and regional diversity) who evade easy classification into consolidated audience segments? How is the approach of global players like Netflix different from that of domestic players like SonyLIV and ALTBalaji when it comes to planning the distribution of their platform content?

Across these questions, I maintain an entwined focus on describing the backend of the distribution system that allows for movement of streaming content into mobile phones and how distribution professionals working at various stages of this distribution infrastructure were keen to reach the new mobile internet users. To begin, I provide an elucidation of my methodological approach to studying streaming distribution systems in India.

Approaching the Streaming Distribution Ecosystem

Internet-distributed video services, also called over-the-top video services, have emerged in India with various business models, including SVoD and AVoD. They have been of interest to scholars for many reasons. Some researchers have studied how streaming audio-video platforms have provided more opportunities for new talented voices, instead of perpetuating the hegemony of stars associated with legacy media like cable TV or Bollywood cinema. Political economy researchers worry about the influence of Big Tech trying to capture the Indian market—which consists of millions of users and is potentially the second largest audience base in the world—and hence renewing concerns about "platform imperialism"[9] This leads to debates about cultural homogenization, but some others discern a difference between cultural imperialism and media imperialism.[10] The Indian streaming ecosystem is witnessing a diversification of domestic players and the entry of global actors, leading to strategies of both competition and collaboration between audiovisual players (Hotstar, SonyLIV) and communication players (Jio, Airtel).[11] This categorization of media industries/platform companies (content production) and communication industries/telecom companies (distribution) is blurred if we think of how some distribution players like the telco Jio run both streaming apps and production studios.

I analyze the different actors and intermediaries involved in distributing streaming content and how, depending on where they are in the streaming life cycle—production, postproduction (publishing content and curating catalogs), encoding and transcoding, and content delivery—they emerge with their own assumed audiences and ways of delivering optimized and convenient services to them. Domestic video streamers in India are reliant on communication and distribution infrastructures owned and operated by other companies. This makes domestic players like ALTBalaji and SonyLIV more dependent on cellular operators and cloud computing companies than some global players like Netflix and Amazon, which have their own content delivery networks, such as Netflix Open Connect and Amazon CloudFront.[12] These aspects of structural dependencies and the role played by Indian state regulators governing platform content and distribution infrastructures are important to consider.

Alongside the political economy concerns of ownership, dependency, and regulation, my focus is on mapping the distribution ecosystems of streaming in India through a method of "tracing" that Joshua Braun has elaborated upon. Braun expands on this methodological approach while describing and analyzing the various "individuals, infrastructures, and services" that helped a news organization make its video clips reach audiences.[13] He explains how a variety of socio-technical systems and actors handled these video clips at different stages of their distribution route. The question for me is this: What does the distribution ecosystem look like for somebody curating video catalogs for easy visibility of on-demand content for audiences? What does the distribution network mean for somebody who is in charge of making sure a high-quality video can be accessed by mobile phone users as they make their evening commute by train or subway, vulnerable as they are to internet bandwidth fluctuations? These kinds of research questions are explored through a methodological approach "to *trace out* the route taken by media products as they make their way from the producer to the consumer, (re)exploring the system from the vantage point of each of the various individuals and intermediaries who handle the content along the way"[14] (my italics for emphasis).

This process of tracing the streaming distribution ecosystem provides legibility to the work of many different intermediaries that make it possible for the public to access cultural content. Such distribution intermediaries are barely accounted for in the scholarship on production studies. The services, infrastructures, and individuals involved in distribution operations—which include encoding, data storage, retrieval, and delivery—and decoding are vast and varied. They exist as transcoding algorithms and storage infrastructures as well as catalog experts and cloud computing architects, and they can be found spread across various sites, from production studios to data centers. In order to examine this streaming distribution ecosystem, I therefore refuse to be limited to studying just one kind of site (studios or data centers) or just algorithms or just infrastructures. In researching streaming distribution systems, I draw inspiration from current work on "media backends" (like algorithms, infrastructures, and distribution systems, in contrast to media frontends like computer and TV screens). Parks et al. have suggested that scholars should move beyond studying "sites of production/consumption" to "studying productive relations across the many sites of supply-chain."[15] Distribution systems are media backends in the sense that such systems seem to be away from "us" as users—physically, mentally, spatially, and temporally—but incessantly enable our interactions/intra-actions with digital interfaces/intrafaces next to us. In studying such distribution systems as media backends, I focus on the relations across multiple sites of streaming media delivery, storage, cataloging, and transcoding.

Such a methodology is necessary, as became evident during my conversation with different intermediaries who are involved at various stages of the media production and distribution process: the SVoD content providers admitted not knowing much about content delivery, and the content distributors wished they could interact more with their content-provider partners so that they could both learn from one another. A content distributor noted in an interview with me that "content providers are used to dealing with the CDN as a black-box. They are used to thinking of CDN as having effectively unlimited capacity for storage," which, this interviewer emphasized, is not always true, especially if one is not a global CDN like Akamai Technologies or Microsoft Azure. This same interviewee also noted that he is oblivious of the technical sophistication involved in encoding and packaging the content before it comes to the servers he configures, caches content for, and maintains for the media delivery. Thus, streaming distribution through the internet is not a single-step process but consists of a delivery chain that involves practices of publication (encoding/transcoding), transmission (carried out by CDNs), and reception (decoding). The purpose of this chapter is thus to draw connections across these multiple sites, backends, and processes.

Finally, it needs to be said that given the multisited-ness of distribution infrastructures, there is no one map that can be traced. Rather, what emerges are "nested maps" of "hard and soft technical infrastructures, open and closed knowledge systems, public and private investments" on distribution infrastructures.[16] This is especially true for a country like India, where there are vast infrastructural inequities as one moves through the disparate geographies from metropolitan cities to tier 2 cities and then further into rural areas.

For this chapter, I draw on about fifteen open-ended, long-form research interviews with people working at various stages of production, postproduction, and distribution in the streaming-platforms industry, as well as fieldwork in and ethnographic observations of the city of Mumbai. In addition to these, I rely on analysis of popular and financial reportage on the streaming media industry, as well as the Indian government's regulatory documents and industrial corporate reports, for the arguments made about the contemporary video-streaming ecosystem in India. In the interviews I conducted, the media distribution professionals at times provided insightful observations about their and their colleagues' work in relation to the current state of the distribution industry, but at other times, some of their comments seemed like a pitch for their CDN or for their company's strategies. Media industry scholars have noted this aspect of interviewing industry insiders, whose insights can be "theoretically suggestive, and culturally revealing" and at the same time be some form of "promotion, and spin."[17] While there is much to be gained from industrial self-reflexivity and distributors' self-theorizing, it is important to contextualize some of their statements and comments in light of other structural concerns and contingent circumstances.

Starting an Audio-Video-Streaming Platform

As I noted in the beginning of this chapter, several former TV channel companies were moving into the OTT/video-streaming space around 2015. Star Network channels had already moved into the space with Hotstar. A former executive of Viacom18, Rahul Vaz, mentioned that this company was also planning to implement a streaming model around this time. As part of his role at Viacom18, Vaz used to help with the online presence of some of the channels, like Nickelodeon, Colors, and MTV. These websites had not been revamped as part of the Web 2.0 platform shift yet. Viacom18 was planning to combine these channels and present their aggregate content on a streaming platform called Voot, which they had planned to run on an AVoD model. Vaz found himself working with some in-house teams who had experience in strategy, technology, and media, and some content and technology teams were also hired from elsewhere.

The plan for Viacom18 was to make the Voot interface look different from the Hotstar interface as a way to project its uniqueness. The platform interface would have to be easy to navigate for the diverse users of Voot watching the different channels. Viacom18 had the following channels: Nickelodeon for children, MTV for the youth, and Colors for families. These different user categories had to find a way to get to their content. As they navigated through the interface of Voot, users would not see their particular TV channel but would see rows of catalogs of content. Could a row be assigned to content catalogs from each specific channel? If MTV, Nickelodeon, and Colors were each going to have either own separate row, which channel would get the top row? Which set of users could be relied on to scroll down a couple of rows to find their favorite channel's content? In making these adjustments, the publishing team was shifting from the rhythm of live television to the updates of the online world. The plan was to refresh the content three times during the day so that the hypothetical user visiting the interface in the morning, afternoon, and evening would see some new content each time. As Vaz put it, "Everyday, I wanted to make a catalog available. What can I present to them which is fresh? Somebody had to come at 7 A.M. and create those rows, [so that] they see fresh content."[18]

The planners of Voot from the Viacom18 in-house team also did not have experience of managing digital content. For this, they sought outside help from Kaltura, and for content delivery, they had Akamai Technologies as the CDN. These tie-ups with CDNs were the norm then, at least for domestic players in India. The OTT services could not simply depend on their own internal data center operations and had to rely on cloud technologies to scale and deliver content to millions of users with the help of a CDN, which is characterized by just-in-time computing methods.[19] To begin with, both Voot and Hotstar went with Akamai Technologies as their third-party CDN, with another backup CDN in case Akamai's servers were down. This backup option was exercised through a manual switch. Even today, Akamai remains their CDN of choice. However, since 2020, a multi-CDN approach has been implemented where, based on performance monitoring, the switch to other CDNs happens automatically. This is a more fluid conception, with multiple CDNs such as Akamai, Google's Cloudflare, and Amazon's CloudFront operating.

The Use Case: The Mobile Phone Audience

After emphasizing this reliance of the Voot streaming platform on Kaltura for video management and on Akamai for content delivery, Vaz went on to explain that—to ensure quality of service for their customers, along with video management and content delivery—they had to optimize the encoding process through the bit rate ladder. In this arrangement, videos are packaged in arrays of segments of different quality and resolution. The source content could be shot in 4K with Dolby sound, but if a user is operating under a bandwidth constraint, then to reach them, the 4K has to be downgraded to HD, or the sound channel has to be controlled so that, out of seven channels of audio, only five are made available. This might lead to the loss of surround sound, but the audio-video will not buffer. There needs to be as little buffering as possible because, as Vaz explained, the "buffering" experience would lead the user to "choose a [downloaded] mp4 file rather than streaming. . . . So the initial battle was that . . . we want people to disconnect from watching an offline, mp4 pirated content."[20]

As Vaz explained the stakes in making the streaming experience effective and seamless, the scenario he picked was that of a user in a commuter train with their mobile phone on 3G or 4G, who at that time would watch a downloaded mp4 file of a film or TV show rather than stream it because of the anticipated buffering experience. To encourage this user to stream rather than watch an illegally downloaded file, buffering had to be reduced as much as possible. Around 2016, when Voot was about to be released, Reliance Jio had launched and was steadily offering 4G data services on phones, but these services would still fluctuate as one was moving on public transport. In the interview, Vaz explained in some detail the user scenario, which needs to be understood in the context of 2016 and the particular kind of mobile phone usage and technology prevalent then:

> Let us assume that now people have 4G access. Even with 4G if I am going in a train in Mumbai there are these blind spots where the network just is not available and this is there even today in a city like Bombay. So at my home it is good, in my office it's good, but my purpose of entertainment is not in office when I am working . . . not at home. . . . Maybe in home I do not have time, unless it is everybody watching something together. It is on [the] journey where it is my niche ["me"] space, right? And in that time if my network is going to fluctuate I may just get irritated, so what I will do is I will just take. . . . Earlier they used to get a movie file mp4 and it was on a China phone or an Android phone. . . . It was easy to just have it. You have one movie at a time, play it back and delete it. What we are saying is we do not want that to happen. We want you to stream. .So, we want you stream at different bitrates . . . and then there are other optimizations. . . . Bitrate is one of them. . . . The other how small chunks do you make? How flexible is the player to adjust? As soon as it knows the network, does it really seek for a very high quality content? If it looks for higher quality content,

> and the player is not flexible, then it will go through that problem. . . . So from a manufacturer's side, I can create ten streams and I am pushing it out . . . but the technology on the player side, like if the player realizes that network has dropped, then next ten packets which are aligned I have to automatically switch to a lower bar. Although it is there, the playback to do that heavy lifting, and if that is not happening, I am going to get an experience of watching content that is buffering, and so it is going to make me choose an mp4 file rather than streaming.[21]

Vaz offered a lengthy explanation of how ten streams of different bit rates with a ladder/incremental approach to video resolution and quality are created for different users watching at different bandwidths, depending upon their mobile phone connectivity as they are moving and the video player (playback) qualities of their mobile devices. The video-watching experience as a private one, outside the office and home, in commuter trains or on auto-rickshaws, is a scenario I heard from several streaming distributors I met during my fieldwork. When they started streaming in 2016, for Voot and for several other streamers, the quality of streaming had to be good enough so that the mobile phone user stopped watching already-downloaded content and felt confident enough to stream without encountering buffering. The transcoding approach of bit rate ladders was a way to overcome the constraints of bandwidth and bottlenecks of delivery.

This idealized case of a private bubble of individualistic viewing for phone users who used to watch predownloaded content has been contested by some scholars, as well as by industry reports suggesting that the trend actually has been toward more social and shared forms of viewing, both on the small screens of smartphones and on the medium and large screens of laptops and television sets.[22] That said, some variation of the same scenario—of a mobile phone user, usually a man, having a private viewing experience on the move—for which the distributor was trying to guarantee a seamless streaming experience persisted throughout my interviews. This trend of subscription services providing mobile apps that allow customers to reconstitute the "viewing experience of the living room in train cars and passenger seats" is something that Ethan Tussey has written about in the context of the United States.[23] The key difference in India, frequently remarked by the distributors I interviewed, is that the commuter experience of mobile phone entertainment is often the only private time these people have, because at home they have to watch the living-room television in a shared family environment. Another crucial variation from Tussey's examples is that, in many cases, the mobile phone is the only screen they have to watch SVoD shows; they simply cannot afford a television. That said, there is definitely a resonance in the Indian streaming distributors' views with what Tussey understands as how US-based streaming services are targeting their mobile phone customers' downtime, and this commuting time certainly becomes a period when the mobile phone user can be made to watch shows. This targeting of the commute brings attention to the spatio-temporalities of mobilities (un)limited, where the mobile phone at once provides users the ability to watch on-demand shows through an on-the-go screen interface and makes available another opportunity for platform players to have audiences hooked on their shows.

This kind of scenario was repeated by Samar Khan, a showrunner who has directed and scripted many TV shows for a variety of OTTs and who is presently the CEO of Juggernaut Productions. I met him at the Andheri West offices of Juggernaut on a rainy Saturday afternoon in August 2022. In his role as the showrunner, he works through various stages of OTT content development, from pitches to collaborative script writing to shooting and postproduction. Khan noted that with mobile internet connectivity being so high in India and the country's mobile internet prices being so low, it is safe to assume that people are using mobiles as their primary source of screen time and that they are watching streaming content on their phones. Khan explained how the length of the show is a key consideration and that the content, according to him, was to be driven by "characters" rather than "stories" or "plot" (narrative) with repeated sensational "high (drama) points" injected/inserted every six minutes. Here is a long excerpt where Khan details the mobile phone audience of streaming shows:

> Number one you have to keep in mind . . . holding up a phone up for thirty to thirty-five minutes is tiring. We are struggling with duration, number one. That you need to make it . . . thirty to thirty-five minutes seems to be the sweet spot of content. . . . ALTBalaji was the platform that was giving fifteen to twenty minutes of content per episode, they were like, do not cross twenty minutes because . . . they were looking at lowest-of-lowest . . . you know, denominator for their audience. They were looking at audiences which were either traveling by trains or in public modes of transport . . . did not have their own cars . . . that was their demographic. . . . So if you were sitting in an auto-rickshaw and watching content, you would want to quickly finish. Plus, you would be shaking hands. . . . You would not like it to be too intense. . . . You do not need large visuals, you need more drama. It has also given more emphasis to characters where people remember characters than stories. . . . You might remember characters of a great TV show;

you might not remember the entire plot. . . . So, more emphasis is being given on characters. In a thirty-minute episode, you are looking [at] five high points. That after every sixth minute, there should be something that pulls me back again.[24]

Khan notes that while some streamers were working with thirty to thirty-five minutes' duration for an OTT series, one streaming platform, ALTBalaji, went for fifteen to twenty minutes in hopes of targeting the "lowest-of-lowest . . . you know, denominator for their audience," who were traveling on "public modes of transport" or auto-rickshaws with shaky mobile screens to watch content on. This way of suggesting the class composition of the newly mobile consumer—the aspirational neomobile, who may not own a car but can afford a newly acquired smartphone—was something I encountered in several interviews, whether explicitly like in Khan's words or more implicitly. This way of describing the undifferentiated, homogenous, and abstracted category of neomobile OTT audiences had connotations of the "masses" of "mass media."[25]

The new-to-mobile-phone audiences were indeed lumped together as "masses" in various media industry discourses, but it was stressed that these neomobile masses wanted individualized viewing of streaming content on their phone screens, as is evident from this statement by Ekta Kapoor, the CEO of ALTBalaji:

"We saw the need for individualized content consumption come into play. And we know that the Netflixes and the Amazons . . . have the money to make high budget large sized productions, which would take care of [the] urban need of individual consumption of content, but the masses out there, very few people realize that they also need individual consumption of content[,] which means that what I see with my family and what I see with my friends is not what I want to see alone . . . I do not care if South Bombay watches my show or not. I am very happy if from Washi to Bihar, everyone has AltBalaji on their phone.[26]

Kapoor here again differentiates her brand of streaming content from global streamers such as Netflix and Amazon Prime Video, by stressing her audience constituents as being beyond the urban areas of "South Bombay" to the "masses out there" in provincial India of states like Bihar, in tier 2 and tier 3 cities, often referred to as de-anglicized "Bharat" audiences instead of "India" audiences.[27] The distribution geographies of streaming platforms are marked by region and class, as ALTBalaji offers much more affordable subscription packages than Netflix and Amazon Prime. Neomobiles seem to have inaugurated the age of the individual viewer who can finally watch things on the phone screen without having to share with the family as was mostly the case with erstwhile television viewing.

The neomobiles themselves are a highly heterogenous group. One key constituent group is college students. This point emerged during my discussion with Danish Khan, executive vice president and business head of the streaming company SonyLIV. He echoed Rahul Vaz, Samar Khan, and Ekta Kapoor, noting that "India is a country of small homes and large families," where "the only personal content consumption happens on phone."[28] He mentioned that SonyLIV's yearly subscription rate of Rs 1,000 was not affordable for everybody, and as part of telecom bundling through Airtel, Jio, and Vodafone Idea, another significant group of people could watch SonyLIV. While this discounted mobile pack brought in less revenue per consumer than the more expensive direct-to-consumer pack, Danish Khan mentioned the benefits. The mobile-pack model reduced acquisition costs and helped increase SonyLIV's footprint. Furthermore, he reasoned that today's students—who have less income right now—are consuming SonyLIV on a subsidized mobile pack but after five or six years will become professionals with more disposable income. If, at that point, they continue liking SonyLIV's content and streaming experience, they will go for a direct-to-consumer pack to be watched on large TV screens:

> People who cannot afford all OTT at full-fledged rate, they get a chance to sample OTT service at a much lower cost, experience it, in a small screen through telco and other bundling services, to an extent, and a percentage will convert to become full subscribers in due course . . . either growth in their financial . . . like students in two to three years will become professionals. Students who have consumed OTT on a telco bundle will move to buy bigger devices and will consume OTT on bigger devices. So, in five to six years, all sixteen [year-olds] will be twenty-two years, and so I think they will expand the direct-to-consumer business of all OTTs . . . as a sampler with a low cap I think telco bundling serves a great thing . . . which is expanding the OTT content market.[29]

Thus, telecom bundling is seen by SonyLIV and other OTTs as a way for young students to sample their content. In five to six years, these same people, now professionals, will take up direct subscriptions from the streamers. The VoD companies have, for now, decided to ride on the distribution infrastructure of cellular operators such as Jio and Airtel. This brings me to the discussion of telecom bundling.

Telecom Bundling

In many media reports discussing bundling, the distribution partnership between video-streaming platforms and telecom companies is hailed as having made content accessible to a large segment of the Indian population. Video operators like Eros Now, Hotstar, Amazon Prime, and ALTBalaji are described in these articles as appreciating how the telecom companies like Jio and Airtel have helped them distribute their content. This new digital/mobile media distribution infrastructure is touted as a more formalized and evolved arrangement than the early days of satellite television in India, when broadcasters had to pay large "carriage fees" to local cable operators to make their content available to TV-owning households.[30] Despite these celebratory pronouncements from the video-streaming industry representatives, political economy of media and communication scholars have suggested that this dependence of VoD operators on telecom players is concerning. This concern is amplified because the telecom landscape in India is an oligopolistic setup (with only a few players operating and calling the shots), where Reliance Jio is increasingly perceived to be the hegemonic, dominant player.[31]

Reliance Jio has always maintained that it is more than just a telecom operator. The company has followed an ecosystem model to grow its platform business, offering a family of apps that include features and facilities like chatting (JioChat), content services (JioCinema, JioTV, the JioSaavn music app), payments (JioMoney), and e-commerce (JioMart).[32] In late 2017, Reliance Jio struck a deal with Balaji Telefilms to have content from their newly opened OTT arm, ALTBalaji, play on JioCinema and JioTV. After acquiring a 25 percent stake in ALTBalaji, Reliance Industries soon bought a 5 percent stake in Eros Now. Eros Now is the digital streaming-video service owned by Eros International, which possesses a vast library of films made in India. This partnership helped Jio have more content for its JioCinema and JioTV apps.[33]

Seeing Jio build up these content partnerships—which helped it acquire more mobile phone service customers who could watch Eros Now and ALTBalaji content on Jio phones and the JioCinema and JioTV apps—other telcos, Airtel and Vodafone, signed deals with Hotstar, Netflix, SonyLiv, and Prime Video. Over the years, Airtel has offered a tiered set of options to its mobile phone customers. Based on how expensive the package is, particular streaming content is made available, sometimes for a month, sometimes for three months, and sometimes for a year. In 2022, Airtel announced its "Xstream Premium" offer, providing access to over fifteen video-streaming services at an introductory price of Rs 149 a month or Rs 1,499 annually. This large bundling strategy was touted as making Airtel Xstream Premium a "streaming super app" catering to smartphone and smart TV users (figure 2.1).

Super apps seem to have a Swiss-Army-knife-like potential to house entire suites of apps, making them convenient for users, who can have one log-in to one app instead of multiple log-ins and pay for one streaming super app instead of making multiple payments for many streaming apps.[34] Adarsh Nair, chief product officer of Airtel India, mentioned these conveniences while describing the qualities of the Airtel Xstream Premium app: "The current OTT ecosystem offers a fragmented experience. Payments is also a difficult problem when one user has multiple subscriptions. . . . One price. One Login. Content from more than 10 platforms on one single app backed by our technology."[35]

In this aggregator/bundling model, are the telcos gaining an upper hand? Researchers would have a better understanding of the situation if they knew a bit more about the implicit agreements and disagreements between content producers and telecom operators about who gets to decide on rights to what content and at what price. This kind of information is indeed difficult to find. At some level, this telecom bundling arrangement seems like a win-win situation, and often the content producers and telecom distributors want the public to know this. VoD players like Hotstar and SonyLIV seem to be carrying on the legacy of major broadcasters—Star TV and Sony Entertainment Television, respectively—and therefore have at their disposal vast content libraries and long experience of catering to Indian audiences. The telecom operators spend billions of dollars on infrastructure such as spectrum, cell antennas, and optical fiber cables, and generally, their main task is to provide pipes for the VoD content to flow through.

While Jio is the exception in that it does not seem to consider itself merely a pipeline for other video platforms, Airtel seems fine with accepting that its primary objective is not to invest in content programming but, rather, to distribute it. As Adarsh Nair put it, "Our OTT partners do what they do best: curate and create the best content. And we do what we do best: distribution and technology."[36] VoD player Hotstar, for its part, returns the compliment: "Airtel has among the widest consumer footprint in India and

Figure 2.1
Airtel Xtream bundling streamers and Xstream play promoting streamer SonyLIV's show *Scam 2003*.

with our expansive content library, comprising rich programming that can cater across age groups, geographies, and interests, there are natural synergies and we are very excited about this partnership"[37]

Unbundling: Collaboration and Competition Between Hotstar and Jio

I want to explore the relationship between Hotstar and the telcos a bit more and, in particular, the dynamics of competition and collaboration between Reliance Jio and Hotstar. Hotstar, unlike most other VoD players, delivers not only nonlinear programming of on-demand TV shows but also live sports and news. Netflix and Amazon Prime in India deliver only TV shows and movies. While SonyLIV and JioPlay offer some sports content, Hotstar has the distinction of offering sports in regional languages and live streaming the wildly popular Indian Premier League cricket tournament, reaching millions of concurrent viewers.[38] During the IPL season or when the Asia Cup cricket tournament is approaching (also live streamed on Disney+Hotstar), all three telecom companies—Airtel, Vodafone Idea, and Jio—offer discounted plans and packages for their phone service customers to watch Disney+Hotstar.[39] In October 2022, Reliance Jio removed the Disney+Hotstar bundle from some of its prepaid customer plans.[40] What prompted this? Some of it has to do with a recent bidding event related to IPL live streaming rights, and another reason could be that Jio (and Reliance Industries) wants to be seen not just as a telecom/infrastructural pipeline for other platforms like Hotstar but as a company offering a suite of its own platforms.

To more comprehensively understand the dynamics of the Hotstar-Jio relationship, a bit more history about both companies is helpful. Hotstar's cable television avatar was Star TV, which was part of a bouquet of Star channels owned in the early nineties by Li Ka-shing of Hong Kong and then by Rupert Murdoch's News Corp. In June 2012, News Corp acquired a 50 percent stake in ESPN Star Sports and renamed it Star Sports. Star Sports, as part of its "glocalization" efforts in India, not only engaged with live coverage of Indian sports like *kabaddi* but significantly invested in live broadcast rights for cricket tournaments.[41] This paid amazing dividends for the Star Sports channel in a cricket-crazy country like India, as well as among audiences in the Indian diaspora and the South Asian subcontinent, including Pakistan and Bangladesh.

Given this history, when Hotstar was launched, cricket tournaments were live streamed on the platform, and every year Hotstar seemed to be breaking its own record for the total number of concurrent users watching IPL on live stream. For Hotstar's digital distribution support partners—Amazon Web Services (AWS) as the cloud computing service and Akamai as content delivery network—the live streaming of IPL tournaments became a way to flaunt their technical prowess, scaling up to cater to ten million concurrent users each night of viewing. One of the promotional stories on Akamai's website starts off, "On 27 May 2018, Akamai and Hotstar created history by delivering the final of the Vivo Indian Premier League 2018 to more than 10.3 million concurrent users. This was the first time ever in the history of the internet that concurrent traffic breached the critical 10 million mark."[42] Supporting ten million concurrent users is not a small feat in the unicast OTT model, which runs over internet protocol, where every stream is a unique copy.[43] In the broadcast model, one copy transmitted through a multicast signal is watched by everybody. In Hotstar's IPL livestream scenario, if there are ten million viewers watching cricket on the platform's app, that means ten million copies are simultaneously being pulled from the data center network.

Some changes in Disney+Hotstar are connected to shifts at Jio and Reliance Industries. The first change was the movement of Uday Shankar, once a key architect and strategist at Disney to possibly Viacom18.[44] The second change, in some ways related to the first, is that in the IPL auction carried out in 2022 by the Board of Control for Cricket in India, Disney+Hotstar won the TV rights for only the next five years. The digital rights were won by Reliance's Viacom18. Live streaming the IPL tournament would be a new thing for Viacom18, and Reliance hopes that more audiences lured by IPL will subscribe to Viacom18's streaming platforms, like JioCinema.[45] Given his experience with Hotstar and IPL, Uday Shankar might be able to rejuvenate Viacom18, which has been an important TV and OTT media player but has never come close to pushing Disney+Hotstar from the top slot. Within the context of these two related changes, Reliance discontinuing some of the Disney+Hotstar offerings in its Jio telco bundles makes sense.

Philippe Bouquillion and Christine Ithurbide characterize the relationship among and between SVoD players and telecommunication operators in India as "coopetition," involving collaboration and competition.[46] In the bundling and unbundling of Disney+Hotstar offerings in its phone data packages, Reliance Jio was at one point (ahead of IPL seasons from 2017 to 2022) interested in collaborating with Hotstar and at another point (post-2022 IPL auction) trying to make sure its Viacom18 streaming services compete with Hotstar. This coopetition between Reliance Jio/Viacom18 and Hotstar had enormous repercussions, because live streaming cricket, the most popular sport in India, is at once a media distribution challenge with a high number of concurrent users and a chance to witness how much live sports are responsible for the spectacular rise of Hotstar. With Disney+Hotstar losing the rights to live stream IPL, the platform did witness a significant decrease in its subscribers in 2023, when the cricket games were streamed in JioCinema (Reliance/Viacom18's platform). To increase its subscriber count, JioCinema strategized to stream cricket IPL for free, and since then, JioCinema has been streaming a variety of cricket tournaments (see figure 2.2).[47] As I have been completing this book, the coopetition drama continues with JioCinema and Disney+Hostar finalizing their merger to form the JioHotStar platform in February 2025, which also signals a collaboration between Viacom18 and Star India.[48]

The telecom bundling and unbundling is, at one level, part of the relationship between the telcos and their customers and, at another level, a partnership between the telcos and the VoD platforms. From the distribution side of things, I have tried to explain how telcos and the SVoDs are reaching out to audiences, particularly mobile phone audiences, through these bundles, which gives us a way to understand the dynamic relations between SVoDs and telcos.

Content Delivery from/at the Edge

Even in the direct-to-consumer business model, where there is a direct subscription (and not a subscription through the telcos), most SVoDs depend on content delivery networks for their content distribution. As I have discussed, two streaming platforms in India, Voot and Disney+Hotstar, have Akamai as their CDN. Netflix is one of the few SVoDs (besides Amazon with AWS and CloudFront) that has its own CDN, Open Connect, through which it moves content closer to consumers by distributing its servers across the world.[49] A CDN typically has a worldwide network of servers that are strategically deployed so that when a user requests content, it is fetched from the server nearest to the user as a way to ensure speed and reliability of delivery.[50] In

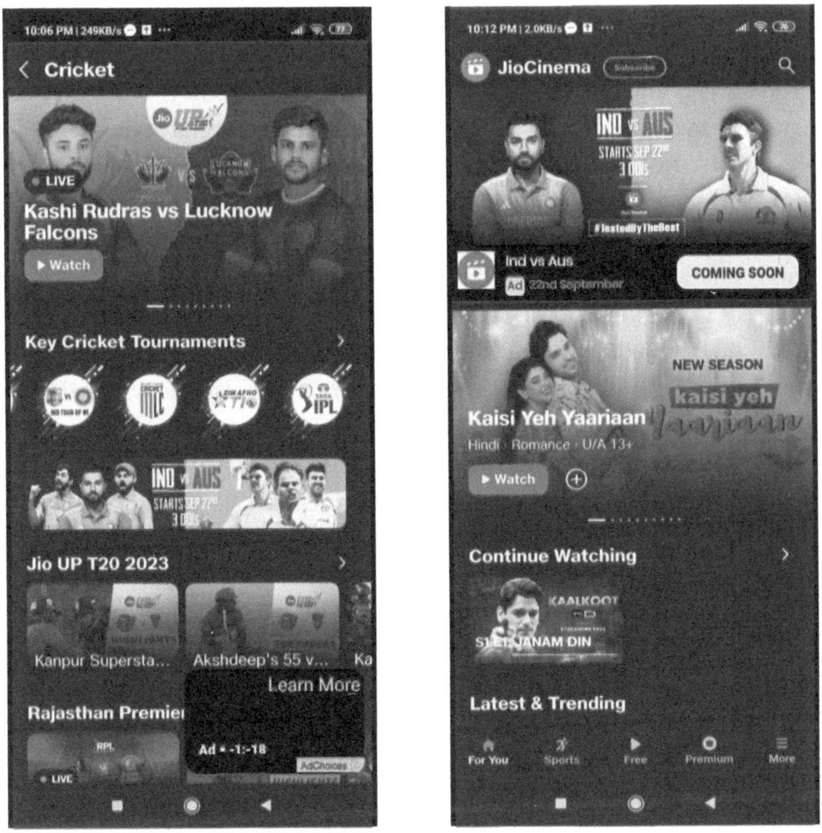

Figure 2.2
JioCinema streaming IPL and other cricket tournaments.

doing so, CDNs reduce the travel distance for media data packets between the source and destination, shifting the source of data to a server closer to the user; this sever is called the edge server. However, such an architecture of media delivery through edge servers also creates the need for CDNs to have local servers in many countries and regions around the world.

Given the scale and flexibility of operations required to cater to a geographically distributed population of customers, the CDN industry is dominated by a few big players like Akamai, Amazon's CloudFront, Limelight Networks, Google's Cloudflare, and Microsoft Azure. While there are small CDNs that do more niche work, there remain barriers to entry because the existing companies have already made commercial investments in maintaining a global spread of servers and have the technical know-how to coordinate data movement across these servers. When a news corporation or SVoD is tying up with a CDN partner, the expectation is often that the agreement will be for media delivery support across several regions around the world.[51] Not surprisingly, these same big CDNs are also the leading content delivery players in India, with Akamai being something of a leader because it has had partnerships with Indian media industry content producers and SVoDs for a longer time than the rest.

There is no doubt that these big CDN companies play a key role in global digital media distribution. That said, given their size, how attentive are they to regional disparities in streaming infrastructures within India? CDNs shape the spatiotemporal logistics of VoD data storage and distribution today, and they can make it very easy for users in certain regions to access particular kinds of content, but they can also exclude users in other regions. In India, a global CDN like Akamai—given the rack size of its assemblage of servers and its expectations of certain minimum traffic speed and concentration (which some allege to be about 8 gigabytes per second [GBPS])—may not connect with a small internet service provider (ISP) in a tier 2 city in India. Even as Akamai and Netflix are committed to distributing content to localized servers, they could create exclusions based on certain bandwidth criteria that ISPs or their preferred middle-class audiences must meet. A comparatively much smaller CDN company representative, who works with local ISPs in tier 2 cities and rural areas of India or small ISPs in hyperlocal pockets of metropolitan areas, had this to say about some of these infrastructural inclusions and exclusions:

> Why does Akamai say that you [local ISP] need to have to have 8 GBPS of traffic before I will talk to you? . . . Because an Akamai node is a rack with 40U of space. . . . At most they will half-populate with 20U. That is their smallest size they can ship to you. . . . That's why, getting an Akamai node into an ISP is a six-month-to-a-year-long process, of starting your initial contact with them, negotiating, figuring out. . . . They do the study of your ASN's traffic pattern, then the physical logistics of shipping such a large number of servers and infrastructure to you, and the same applies whether you are talking to Akamai or to Google, Netflix, Facebook. . . . All of them have their own process, and it is a lengthy, involved, and pretty hardware heavy process. . . . You cannot be having that conversation with a small ISP. . . . He does not have the bandwidth to engage with you. . . . He does not have the number of subscribers. . . . He does not have the internal team which can make sense.[52]

The logistical and technical challenges in the interactions between Akamai/Netflix and the small, local ISP described here already seem to create infrastructural exclusions, even though Akamai and Netflix have a large number of servers/points of presence (PoPs) spread across the world. Concerns about smaller ISPs not being able to negotiate with big CDN companies on equal terms, and about customers of particular VoD streamers receiving preferential treatment because of better CDN services, have been raised in Indian newspapers.[53] These concerns are similar to net neutrality debates that have unfolded in several countries.[54] The Indian government's regulatory authorities have so far been hesitant to put too many restrictions on global CDN operations in India because these companies are believed to contribute to maintaining growth in the overall streaming industry and are offering services at reasonable rates to both SVoDs and customers.[55]

Local ISPs are independent entities that manage the last mile and understand the local nuances. A particular CDN company that works with local ISPs with a smaller coverage zone and limited number of users (from five thousand to twenty thousand), has been propounding the idea of "edge CDNs."[56] I asked what is so new about the emphasis on edge computing, when Akamai and other CDNs also say that their whole delivery process is predicated on the idea of caching content closer to users and their ISPs—that is, temporarily moving whole blocks of relevant content to the network edge. This is an excerpt from Akamai's promotional material (italics added): "Akamai has an unmatched global network capacity of 300+ Tbps and is unparalleled at scale with over 4,200 locations and upwards of 1,400 networks that span 135 countries. With the *largest edge delivery platform*, we see more of what's happening on the internet. This means we can deftly avoid bottlenecks and defend at the edge."[57]

So, if edge delivery is what global CDNs like Akamai do anyway, how are the smaller CDNs special? To answer this, my interlocutor mentioned that "edge" means different things to a content provider, to a global CDN, and to a local ISP:

> Edge is a very difficult word to define because edge means different things to different people. To a content provider, the CDN is the edge; it is the edge of their world. To a top-tier global CDN, you know, who [is] maybe looking at global delivery as their perspectives—a node in Mumbai, Delhi, Chennai covers India as a country—that is their edge. Chennai is an edge node for them. For example, for a CDN like CloudFront, they have about mostly metro location. . . . They have

> PoPs in metros. So, if I am watching from Shadnagar, I am getting served from Hyderabad 50 percent of times (that is my closest PoP) or *addha* [half], 50 percent of time from Bangalore—that is almost five hundred kilometers. How they decide to serve me from Hyderabad half the time or Bangalore half the time, it is a black box as far as the OTT is concerned. The content provider has no visibility on it. I as a subscriber have no visibility on that. A telco has a very different perspective on that. A small ISP has a different perspective on what an edge is. A large, metro, broadband, fiber ISP has a different perspective. Sometimes when we go there, half the challenge is to change the perspective on what the edge is. We walk into a large fiber ISP. They think, oh, edge servers, they are sitting in our central lock. Our definition of edge by default will never match from the definition of edge from whoever that we are talking to.[58]

Through this example, this CDN company that works with local ISPs at a much more granular level than global CDNs is suggesting that the perspective of being close to a user—that is, being an edge server—can shift dramatically based on what kind of stakeholder one is in the content delivery system. Shadnagar is in the Telangana state of India and is sixty kilometers from Hyderabad, the state capital. It is also about 520 kilometers from Bangalore. These two big cities are where a global CDN like Amazon's CloudFront has its edge servers, which serve VoD content to users in Shadnagar.

PicoNETS is looking to areas like Shadnagar, where you have connectivity but the media consumption experience is suboptimal. An ISP operating in Shadnagar cannot think of colocating an Akamai node, because Akamai would entertain them only if the ISP is at 6 to 8 GBPS of traffic to Akamai, which means an ISP of less than two hundred thousand subscribers will just not qualify. A tier 3 city ISP serving five thousand subscribers will not be able to attract these infrastructure providers, and these areas are where picoNETS is thinking of asking SVoDs to focus, with their help.

PicoNETS is working with much smaller servers programmed at the edge of a local ISP (see figures 2.3 and 2.4). Such a server is configured to be able to anticipate the more popular content watched by people in that area and to load its servers with that content in off-peak hours so that less pressure is created on the backhaul network connecting the local ISP to the internet exchange. Peak hours of video traffic are typically in the evenings from 7 to 10 P.M., and an off-peak hour would be something like 1 to 4 A.M. This is how Prakash Advani of picoNETS explained the spatiotemporal logic of the delivery architecture:

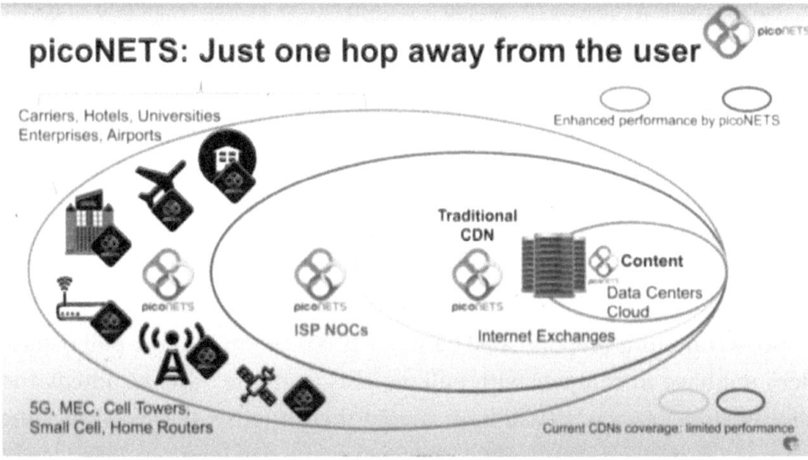

Figure 2.3
picoNETS cache servers located at deep edge close to user devices. Courtesy: Prakash Advani.

Figure 2.4
Rajeev RK and Prakash Advani holding picoCDN cache servers (Andheri Starbucks, Mumbai). Photo taken by author.

> Pulling content between 12 A.M. and 6 A.M.—in fact, as a CDN, we are encouraged to do that. As a CDN we provide the framework. It is how the content provider wants to use the infrastructure to pre-populate the content before the traffic. The ISPs where we locate our node, they know that in these lean traffic periods, their bandwidth is sitting unutilized. Now, if I download this content at 3 A.M. when it is just coming from the transcode pipeline of the content provider into the node, that means at 7 A.M. when a hundred people are looking to watch it, he does not have to go and pull it at that time. Or, let us say at 9 A.M. when that pull of that content is going to be competing with everyone who has just reached the office and is checking their news, everyone [who] is getting on to their morning work and getting on Zoom.[59]

So, if content is prepopulated between 12 A.M. and 6 A.M., the pull request does not have to compete with pull requests from the other hundreds and thousands of users who will have started their internet television viewing from 7 A.M. onward, gradually picking up as the day goes on. This idea of prepopulating the early morning content that is actually going to be shown in the evening is a fascinating (and somewhat counterintuitive) idea and has its own rationale.

The traditional cache model in CDN systems involves trying to understand what is familiar and what is popular in a given location. Based on such a data monitoring exercise over time, the CDN decides to off-load that particular content into the specific local network server. Rajeev RK, the chief technological officer of picoNETS, finds two problems with this traditional model of caching.[60] First, it takes time for the cache node to see traffic patterns and identify what is popular, which means the early viewers are going to get a suboptimal experience until the node figures out if this content is popular and should be locally stored at the edge server. Second, a streaming show's early viewers are probably their most important customers because they are the ones who are driving trends. But, because of the mechanics of the cache model, the SVoD might risk losing them. Keeping in mind these eager audiences and fans of the show, SVoDs, on Rajeev's advice, have decided to have TV show episodes that are playing in the evening during the week drop on the OTT app at 6 A.M. Along with the dedicated followers, commuters, those people in movement, are in this way captured by the SVoDs because they can watch the show on their way to work in the morning.

The servers that picoNETS operates are caching devices, and because they are operated as a CDN, the company can make relationships with the content providers that allow picoNETS to integrate with the provider's content management system and obtain access to the kind of https certificates required to be able to serve their content. According to picoNETS executives, the providers are also realizing that the large metros are getting saturated. Everyone who is likely to buy a streaming subscription probably has one; most of them may even have multiple subscriptions. Just waiting for the young population to grow up—so that when they become salaried professionals they will eventually start buying subscriptions—is not considered a proactive approach. The only way for the VoD companies to expand their market and grow fast would be to have regional-language content and draw customers from the tier 2 and tier 3 rural locales.

That said, in these semiurban and rural markets, SVoDs have a much harder time because the users are much more value-conscious. Such users in a tier 3 town are not going to pick up a subscription if the streaming service is unable to deliver a good viewing experience to them. The SVoDs and their partner CDNs will need to collaboratively provide sound delivery capabilities that work with mid-tier devices, which in turn would require a shift in existing CDN standards and a greater probing footprint. H.265 decoder offers better compression efficiency than H.264, but H.265 requires more processing power potentially impacting performance on older or less powerful devices. CDNs operating in India catering to neomobiles who often are working with a phone handed down to them from an elder sibling cannot be expected to have top-of-the-line H265 decoder. As Rajeev RK of picoNETS put it to me,

> You cannot just build an app that works great on a flagship, but does not work on their twelve-thousand-rupee Android, which may not have a top-of-the-line H265 decoder . . . which means you have to do H264, which is the hardware available in these mid-tier devices. How do you ensure that someone who has a device that is handed down from his *badhe bhai* [elder brother]. . . . It is a four-year-old Android device; it is going to be H264 only. H264 needs more bandwidth for the same quality. How are you going to give them a good experience when he is using your streaming service. . . . Platforms need to be able to solve them, to crack that market. . . . That is where we are hopeful that the evolution of CDN standards . . . deeper penetration of CDN networks, will help answer that side of the puzzle.[61]

Finally, amid the discourses of globalization and localization, what is "local" for one streamer/CDN might be "national" for another. So, for Netflix, operating in so many countries, a node in Mumbai will be a local edge node, but for a Desi/Indian SVoD trying to tap into regional-language audiences in India and expand to peri-urban and rural areas, the notion of the edge server would be different. These notions of "local" and "edge server" also shape the way delivery is optimized in the distribution infrastructure. I will try to illustrate this with some hypothetical examples, which were provided to me by some Netflix engineers and CDN technical executives.

To begin with one such example, Netflix is famous for having very detailed digital assets for each of its curated catalog entries related to a film or TV show. This includes multiple artworks related to a show, dubbing done in about ten languages, and subtitles in thirty languages, as well as other provisions for aspect ratio, sound, trailers, and other metadata. This all requires storage, and Netflix—having its own CDN, Open Connect—maintains these assets in Open Connect Appliances (OCAs), which it deploys in the networks of ISPs and telcos. From the launch of Open Connect in 2011 until 2021, Netflix developed and deployed "over 14,000 OCAs across 142 countries."[62] While these metadata and digital assets increase storage dimensions for Netflix, they also help it to gather more data to provide a customized user experience: which particular artwork snapshot the user is viewing more, which subtitle language the user prefers, and so on. Even then, to save space in its "local" CDN in Mumbai or Chennai for a country like India, Netflix will not store all the subtitles and related digital assets. So, for *Stranger Things* or *Sacred Games*, as a Netflix engineer put it to me, "we will not probably land in India subtitles in Korean and Spanish."

PicoNETS, working with a more granular idea of the local and the edge server refers to what they are working with as "hyperlocal" or "deep-edge" servers/computing. Rajeev RK outlined a hypothetical example: Within Mumbai, say an SVoD like Hotstar is going to release a show that is dubbed in eight languages (indeed, Hotstar does dubbing in eight languages for all their originals).

> I want the video content to be cached in both Sion as well as in Borivali, but in Sion, I want to cache the video and the Tamil audio stream, and in Borivali, I want to cache the video and the Gujarati audio stream. Now, once you can start becoming granular like that in terms of deciding how to push your content out into your network. That is where you can start thinking, okay, let me be intelligent about how I use these small nodes/ISPs. You do not have a choice, you cannot just push everything.[63]

Sion is a neighborhood in Mumbai, and Borivali is one of the many suburbs of Mumbai. Here, the local—instead of being the whole of India or whole of Mumbai—has become a neighborhood and a suburb. Sion has a significant Tamil population, and Borivali has Gujarat-speaking people. So, in the hyperlocal move of targeting audiences and economizing server space, Rajeev is suggesting it would make sense to cache video for the new episode of a Hotstar original series in the small edge servers with a local ISP in Sion and Borivali. We have to keep in mind that picoNETS is working with much smaller ISPs and cache servers; their server storage space cannot be compared to Akamai's assemblages or Netflix's OCAs. So, after caching the episode in Sion and Borivali, there might not be enough space for audio streams of all eight languages on those servers, and there might be a limit of four language dubs for each show. In that scenario, it would make sense to have English-, Marathi-, and Hindi-language audio streams in both Sion and Borivali, as well as the Tamil audio stream for Sion and the Guajarati audio stream for Borivali.

CDNs work to optimize content transmission paths—hence the calculations about which selections of subtitles and dubs to store. That said, based on the scale of their operations and the storage space of their deep-edge servers, the CDNs emerge with specific strategies. Both Netflix/Open Connect and Hotstar/picoNETS are not doing everything in real time. Based on some knowledge of what languages work where, they are both precomputing possible subtitle and audio-stream requests and prepositioning particular digital assets in specific deep-edge servers associated with a film or TV show.

In addition to the traditional cache model of off-loading particular content into specific local servers, picoNETS and other CDNs offer precomputing models. Whether it involves precomputing or postcomputing decisions, Netflix is in a particularly enviable position because of its ability to have data throughout the delivery chain. As a CDN professional put it to me, "Netflix is the production house producing the content. They are running the entire infrastructure to package the content. They are running the infrastructure through Open Connect to deliver the content. They have this visibility most other providers don't." This ability to collect, integrate, and relate data at every stage of the supply chain helps Netflix with data analytics and microtargeting audiences.

Local CDNs and ISPs foregrounded how crucial controlling the distribution supply chain is in order to ascertain audience viewing habit data, but at the same time the reliance on precomputation suggests that some of their decision-making is already guided by being aware of taste cultures based on local knowledge and intuition about neighborhoods. The CDN/ISP control over dubs/subtitles is reflective of cable operators in particular locations in India removing or adding channels based on popular demand, with local customer feedback provided through telephone calls.[64] While global CDNs work with bigger internet service providers (ISPs) in India that are often telecom companies such as Reliance Jio and Airtel, local CDNs in India team up with independent last-mile operators, many of whom were erstwhile *cable-wallas* (cable operators) who understand local nuances and have long remained connected with their neighborhoods[65] "Addressability," that is, "the ability to" air/stream "different ads in the same ad spot to different viewers/households" based on their specific identity and consumption habits, remained a key goal for distributors, but the path to achieving that continued to be a mix of formal and informal data gathering practices.[66] An earlier moment in the cable TV industry in India negotiated the conditional-access system (CAS) when broadcasters and advertisers sought numerical clarity about the relationship between cable operators and their audiences/subscribers, which had often seemed steeped in informal socialities and fuzzy numbers.[67] This movement across formal and informal data and calculations indicates the need to historicize present streaming data analysis within longer histories of televisual distribution in India.

Conclusion

Since 2016, buoyed by increasingly affordable data-use rates on smartphones as well as the global rise of VoD services, there has been a definitive shift toward consumption of streamed audio and video in India. This shift was hastened by the COVID-19 pandemic, as the anxiety of in-person collective spectatorship in cinemas and multiplexes was replaced by the convenience of streaming content on television and mobile phone apps for at-home consumers.[68] This demand for streamed content is sustained and accelerated by a crowded space of global and local streaming companies in the country, such as Netflix, Amazon Prime Video, SonyLIV, Disney+Hotstar, and ALTBalaji, which offer VoD services through a variety of subscription-based and advertising-based business models. Based on interviews with producers and distributors of streaming content services during my fieldwork in Mumbai, I have examined in this chapter the shifts in both storytelling practices and delivery methods to argue that production practices shape and are shaped by distribution logics. Production cultures and distribution infrastructures are indeed different things to study, but they also share connections, in terms of how they construct the audience, even historically, when it comes to Bollywood films in India. As Tejaswini Ganti writes, "In the case of the Hindi film industry, whether a film even gets made can also be dependent upon distributors and their particular arrangements with producers," which involves speculations about themes/plots and mapping film-viewing preferences in a culturally diverse country like India.[69]

Content creators at production studios, script writers working in writers' rooms, and content heads of streaming services emphasized learning to adapt to changes, such as the deployment of limited-series formats (instead of broadcast TV shows, which run at times to three hundred episodes) and the introduction of "cold opens" and strategic placements of "cliff-hangers" to hold the attention of "new" audiences.[70] Distributors of streaming content in Mumbai described their neomobile audiences as watching shows on handheld mobile phones in commuter trains and auto-rickshaws, which required tweaks in distribution logistics, including incremental bit rate ladders and deep-edge computing, to make sure there were not many disruptions in viewing. Such a streaming experience was necessary, as distributors believed that if there were glitches or buffering interruptions, then the audience might shift to a different streaming service or app.[71] Thus, producers and distributors construct audience/user interaction scenarios (across caste, gender, and regional differences) in similar ways. This new audience is imagined as flexible, mobile, exercising media entertainment choices, and having a short attention span. In the next chapter, about short-video apps, we will again be discussing how neomobile audiences are being targeted by these vertical video platforms. In fact, one could even consider in what ways short video content has affected the considerations of streamers who rely on longer narratives. During the final stages of writing this book, in May 2025, there is a splash of news stories about microdramas hitting the Indian market. These are vertical video and reel-sized snippety two-minute-long episodes, with about fifty such episodes comprising the whole series. The first five to ten episodes are made freely available, after which the user pays about five rupees for each episode. Local language audio content provider KukuFM has transformed itself into vertical video storyteller, KukuTV, and others from both long-form streaming platforms and short video apps are joining the race to produce microdramas, having known about their success in China since 2020.[72]

Changes to the aesthetics of content with an imagined mobile phone audience watching on a small screen in the video producer and distributor's mind is certainly not confined to India. Tori Omega Arthur explains that the drama series of the subscription streaming service iROKOtv (operating in Nigeria and the wider African continent and diaspora) involve frequent sequences of medium shots and close-ups of characters that make iROKOtv viewers focus on characters' gestures and facial expressions, and such a stylistic choice does not "inhibit viewing" on a "three-by-five inch mobile phone screen."[73]

This chapter has also detailed the various stages of distribution. Over time, content streaming companies have increasingly relied on third parties to provide them with content delivery facilities as well as colocation data centers. While this frees the streamers from having to bother about distribution considerations and hence to focus their energies on developing better content, some scholars have expressed concern about this loss of control over distribution. Over time, most content creators, whether news broadcasters or streamers, have outsourced CDN deployment. That way, the content producers do not have to worry about monetary investments to match their distribution with the demand of content. With an already large number of servers under their control worldwide, global CDN companies have no issue scaling up and down. As Chalaby and Plunkett put it, "They [media owners] can upscale without capital investment because expenditure increases only as needs expand. And for the CDN providers spreading their own investment and cost across thousands of users, this expenditure represents a fraction of that involved in running a proprietary CDN. Being able to leverage scale, large CDN providers like Akamai are profitable."[74] Scaling up server capacity at short notice is possible for Akamai but not for streamers like Hotstar or SonyLIV.

SonyLIV's CEO, Danish Khan, stressed that his streaming company did not believe they can have control over distribution technologies. He differentiated his content-based company from a "technology company," which, according to him, was the way to characterize distribution companies. SonyLIV focuses on making better content and then pairs up with distribution companies like YouTube, AWS, and Jio to reach audiences: "Sony does not have its own content distribution network. We use Jio, Akamai, and AWS to distribute. We personally believe we do not have to own all technology. We strongly believe we are not a technology company. We are a content company. We will work with the best technology companies. We have the world's second-largest YouTube channel, which gives us huge reach and long tail monetization opportunity. . . . We have worked with Jio, Airtel, Vi, and many ISPs."[75]

Not all CDNs are big and global. Local CDNs like picoNETS emphasize the need to create storage and distribution infrastructures that can work with local internet service providers who sometimes do not qualify for global CDNs' devices and equipment, given limitations of rack size and traffic speed/bandwidth. Local CDNs stress the importance of hyperlocal deep-edge computing, where, anticipating specific demand, smaller servers are placed much closer to user devices so that the backhaul network is more optimally utilized. Since deep-edge computing is working with data away from data centers (the "cloud") and closer to users, this process, thought of in environmental metaphors, is also called "mist" or "fog" computing (as if data condensation is happening closer to the ground, or user).[76]

More than just technical operations, the use cases anticipated by local CDNs indicate how streaming distributors take into account India's varied bandwidth geographies (across tier 1, tier 2, and tier 3 cities/towns) and linguistic/regional diversities as challenges to their distribution operations. These scenarios include calculations about which selections of subtitles and dubs to store as CDNs precompute possible subtitle and audio-stream requests and pre-position particular digital assets in specific deep-edge servers. The curatorial work within these CDNs/ISPs draws from an earlier moment in television history, where small local cable network operators, or *cable-wallas*, removed or added channels from their television network lineups in response to customer requests coming to them in person or over the phone.

A history of media distribution in India will connect the churn currently happening in media industries because of the entry of new mobile phone audiences with other moments of disruptions in the distribution infrastructure. For example, in the 1980s, the emergence of video parlors where videotapes were watched on television precipitated a crisis of audience recognition for the more formal Bollywood cinema industry and the state-owned television broadcaster Doordarshan.[77] With the emergence of video, and then the concomitant rise of VCRs, upper-middle-class audiences retreated to home entertainment options and stopped frequenting movie halls until the multiplex phenomena of early 2000s lured them back to public viewing experiences.[78] Since the late 1980s, an informal network of small-budget Hindi films, often referred to as B- or C-grade films, which were categorized as catering to the "lowest denominator," appeared in single-screen theaters that, having lost their upper-middle-class audiences in 1980s, were exhibiting films for the lower middle class and/or working class.[79]

As discussed in this chapter, Samar Khan mentions over-the-top (OTT) player ALTBalaji as making streaming content available for the "lowest denominator" of audience, who have finally, with the mobile phone, found a way of consuming such content in individualized viewing spaces. In this admission, there is an implicit historical gesture that the B- or C-grade movie audiences of the now-closed single-theater cinema halls have been brought back, thanks to the streaming distribution infrastructure, in the form of neomobile streaming audiences of ALTBalaji. This is a broad-stroke history with several reductive generalizations but is one way of understanding the mutating relations between audience imaginaries and distribution infrastructures. More empirical research is needed to scrutinize this historical interpretation. While this chapter has discussed long-video-format content, the next chapter focuses on the "spreadability" of short video content—the ways in which microscale content circulates at macroscales.

3 Spreading: Viral WhatsApp and Post-TikTok Futures

This chapter maps the shifting app ecosystems in India, particularly those dealing with social media, short videos, and instant messaging services of the last ten years. At one level, this chapter examines the media practices, networked affects, and platform affordances that influence pathways of content distribution, especially user-generated content, across the app ecosystem. At another level, the chapter describes and analyzes how apps and platform companies have entered the Indian app ecosystem at different stages and have been competing with one another for garnering the attention of Indian mobile phone users, especially targeting those users who have newly acquired the mobile phone. This is part of these companies' efforts to make inroads into the next two hundred million users in the hinterlands and provincial regions of the country.

In 2020, tech journalist Pankaj Mishra, who started *FactorDaily*, a long-form news publication that reports on the intersections of technology and society from Bengaluru (Bangalore), told me that "the first election campaign that Modi did was a Facebook election. The second election was a WhatsApp election. And if things were going [fine], this would have been a TikTok election."[1] What Mishra is alluding to is that the 2014 election was one of the first elections in which social media played a big role. A significant step taken by Narendra Modi and his political party (Bharatiya Janata Party [BJP]) was to use Facebook for political advertising as well as for political rallies and speeches, which were livestreamed on Facebook Live.[2] Modi was victorious in 2014, and in 2019 when he ran for reelection, the platform of choice was the instant messaging service WhatsApp, where the BJP's information technology (IT) cell targeted mobile phone users through WhatsApp messages. These messages were meticulously designed with "specific cultural and linguistic 'tells' of the region" where each mobile phone user lived, based on the data that the party had previously collected about users through their membership drives and the "missed call" initiative.[3] Mobile phone users then shared these WhatsApp messages with their nearby circle of friends and relatives, which increased the messages' believability even if some of them were propaganda and misinformation. Mishra suggested that if the Indo-China territorial disputes had not begun, and if TikTok had not been banned in India in late June 2020, then it would have been the preferred platform of election propaganda for the 2024 election. TikTok's popularity, especially among audiences and users in tier 2 and tier 3 cities, made it a key platform for political parties to target vote-banks in provincial India. However, the platform faced the ire (and politics) of the Indian state because geopolitical struggles increasingly shape governmental actions regarding digital platforms in the "postglobal" era.[4]

It goes without saying that political parties use a mix of different platforms for their campaigns, but the platform of choice is also an important thing to consider. That the platform of choice was once a social media platform like Facebook, at another time an instant messaging service such as WhatsApp, and was predicted to be a short-video platform (TikTok) in the near future should intrigue scholars of media distribution. In studying platform sociality and "spreadability," I argue for a comparative rubric that goes beyond comparing the memetic quality of merely two short-video platforms or the viral affordances of just two instant messaging services. After all, YouTube video links flagrantly move through the WhatsApp messaging service, and Indian homegrown vernacular short video platforms like Moj make it very easy to share videos not only within the app but also through WhatsApp. The comparative rubric therefore has to concentrate on tracing relations across an ecosystem of apps. Furthermore, most books about media distribution focus on the nexus of the media industry and distribution. This chapter is also interested in how corporate power controls distribution logistics and the way that it shapes the spaces where, and times when, media become available for consumption.[5] That said, while I do focus on the media industry dimensions of content distribution, I remain interested in exploring the circulation of user-generated content and political advertising as well because they, too, involve precise targeting of audiences/markets/consumers, and they, too, involve how media move through our digital lives.[6]

While considering the spreadability of user-generated content in India, at times a particular instant messaging app service like WhatsApp has been singled out as the app responsible for lending virality to rumors and misinformation through forwarded messages. However, social media apps such as Facebook and short video platforms like TikTok, and for that matter, the most durable video-sharing platform, YouTube, have also afforded the opportunities and architectures for memes to spread. While viral posts and memetic videos are one way to trace the app ecosystem in India, they are not comprehensive enough to understand the more habitual exercises of sociality involved in sharing posts and videos among friends and relatives as well as the everyday performances of content creators in making and uploading videos, tweets, and reels, not all of which necessarily become viral. These everyday performances and quotidian microhabits of liking or forwarding a video do have a role to play in making videos spread—that is, when they do spread.

This chapter on the shifting app ecosystem is therefore about the varying socialities and the way they scale spatially and temporally, both up and down, as users interact with content and with one another through apps. Today, sociality is "enjoyed and exercised" through "commercial platforms," and this "platformed sociality"[7] is also a "scalable sociality"[8] given the networked connectivity of platforms that spread content and enable social interactions across transnational scales. In studying socialities and spreadability, I attend to both the quotidian dimensions of individual and collective practices (and feelings) of ordinary social media users as well as the organizational and political economy dimensions of macro- and microdecisions made by app companies, professional intermediaries, and influencer marketers in their efforts to acquire and professionalize particular content creators, push specific content, and consolidate users. This consolidation of users, happens by making them spend time on the apps viewing the apps' contents and promotional campaigns. Since the TikTok ban in India, short video platforms like Moj and Josh have devised strategies to target audiences in tier 2 and tier 3 cities. Josh dispatched mobile stages on canter vehicles into rural India to garner the participation of regional microinfluencers, with its CEO Umang Bedi claiming that Josh was better positioned to understand the "real" authentic "Bharat" India of regional towns and villages than multinational companies such as Facebook and Google.[9]

The chapter investigates how both individual media user's practices and the decisions of platform companies regarding app interfaces, machine learning algorithms, talent acquisitions, and content choices have impacted the journey of information and videos through Indian public cultures. This dual technique of studying the app ecosystem is able to focus on platform-specific features of one app that instigate/afford a particular media-sharing practice and also understand that platform sociality and economy take shape in the complex ecology of various apps, where competitive, adversarial, imitative, and collaborative relations across platform companies (and third-party services) are important to consider. Not just conceptually but also empirically, such a research move is significant because the Indian app scene has witnessed, at various stages, contestations between Western Big Tech apps like Instagram and WhatsApp (Facebook/Meta) and YouTube (Google/Alphabet), on one hand, and Chinese apps (TikTok, Kwai, and Helo) and vernacular Indian apps (ShareChat, Moj, and Josh), on the other.

In what follows, I work through some of the scholarly debates regarding viral media and spreadable media in the field of digital media studies and contextualize them in relation to the history of short-form video in India as well as some earlier research on the proliferation of memetic digital texts in the Indian media landscape. I then move on to discuss how particular cultural dimensions of Indian mobile phone users are entangled with the affordances of particular app features to create situations in which apps like WhatsApp and TikTok become conduits for the virality of messages, information, and rumor. WhatsApp is at once the most common app for everyday communication as well as the app singled out for causing the most vicious misinformation campaigns in India. Carefully attending to the forwarding practices of WhatsApp users in India, as well as the distribution ecology of WhatsApp groups, I explore the nuances of the above-mentioned paradox.

If in the case of WhatsApp, it is the forwarding practices through semi-anonymous groups that leads to virality; for TikTok, it is the video-editing and duet features embedded in the platform design that foster the proliferation of memetic texts. It is important to examine aspects of sociality and spreadability not just through studies of WhatsApp and TikTok but also through the wider ecosystem of apps. I explore the concomitant rise of indigenously designed vernacular apps and Chinese platforms in India since the more widespread availability of mobile internet (also termed the "Jio effect"), especially from 2018 onward.[10] Based on conversations with app developers, user experience designers, venture capitalists, tech journalists, and app users in Delhi, Kanpur, Mumbai, and Bengaluru, I focus on several features of these apps that made them popular among the new segment of regional users in India (often referred to as "neomobiles"), many of whom communicate in and understand only non-English vernacular Indian languages like Hindi, Bengali, Kannada, and Tamil. In particular, I pay close attention to how such Chinese apps, through their algorithms and recruiting teams that selected local performers, were able to hyper-localize their app's following in India (like India's vernacular apps) and become key distribution channels for audiovisual content. TikTok, even though banned in India amid rising escalations of tension on the Chinese-Indian border in June 29, 2020, was the most popularly rising app of 2019 in India.

Even after TikTok left India following the ban, there have been efforts to clone TikTok interfaces and to build similar recommendation systems by home-grown Indian apps like Moj and Josh so as to onboard successful content creators and influencers who had been active on TikTok. Even more specifically, talent acquisition drives led by Moj and Josh focused on garnering hyperlocal microinfluencers in tier 2 and tier 3 cities and towns. This emphasis on "vernacular creativity" as being crucial to tap into audiences in provincial "Bharat" hinterlands, differentiated from the metropolitan India of tier 1 cities, has been an important legacy left by TikTok.[11] In following talent scouts on their recruitment drives and interviewing influencers/content creators in the short video platform industry, I remain committed to studying the "role of human agents (other than those at the pinnacle of conglomerate hierarchies) in interpreting, focusing, and redirecting economic forces that provide for complexity and contradiction within media industries" as importantly stated by Timothy Havens, Amanda Lotz, and Serra Tinic in their influential essay explaining "critical media industry studies" as a research approach.[12] I do perform the "midlevel fieldwork in industry analyses," which accounts for the complex interactions among cultural, technical, and economic forces, thereby appreciating the interplay between structure and agency.[13]

Spreadable Media and a Brief History of Memes and Short Videos in India

In *Spreadable Media*, Henry Jenkins and his cowriters have influentially argued that "virality" and phrases like "viral video" and "viral media" tend to compare systems of "cultural distribution" to "biological systems."[14]Jenkins argues that the easy sliding of "cultural material" into "biological processes" in the viral media thesis inadequately accounts for the participation and/

or effort that the user/audience puts into spreading the video. Some forms that this participation can take include making mashup videos and adding subtitles to the already spreading video. In some ways, the spreadable video is spreading, but it is not just spreading in the same form. Some other audiences, fans, or users are also doing format conversions, and still others are thinking of prioritizing the particular platform on which to circulate the video, so that it spreads fast or, put into another epistemological paradigm, becomes viral.

Some digital media studies scholars, like Limor Shifman, have made a distinction between "virals" and "memes": "Whereas the viral comprises a single cultural unit (such as a video, photo, or joke) that propagates in many copies, an Internet meme is always a collection of texts."[15] For Shifman, the key difference between internet memes and virals is variability: For a meme to be a meme, it should vary in each replication, and so for each new reiteration, there is a change.[16] While the meme can be differentiated from the viral, even a meme is biologically derived, with Richard Dawkins suggesting that memes are cultural parallels to biological genes, in the sense that they self-replicate and mutations accompany that replication.[17] Beyond objecting to the biological associations, Jenkins argues against descriptions like audiences/voters getting "infected" by the affective contagion of corporate marketing and political rhetoric in the viral media thesis because such descriptions overestimate the power of corporate media companies and political propagandists and underestimate the agency of audiences. Jenkins' work on fan cultures and convergence media is legendary, and yet it remains to be seen, I would argue, whether all audiences are always as active as "fans" or if some of them can be somewhat impulsive and do indeed get affected (if not infected) to imitate other users in their community, who are themselves imitatively spreading media content immediately and quickly.

I say this because I did meet people who circulated/forwarded/shared rather brutal videos even before they had watched two minutes of them, through their mobile phones and WhatsApp messaging services. Most certainly, some of them were active audiences in the Jenkian sense, in that they actively manipulated images and subtitles to distort the context of communal riots. So, I did encounter active audiences and I also encountered audiences who were impulsive, infected, and caught in the affective contagion circuit of viral videos. Another thing that somewhat undermines Jenkin's overemphasis on creative participation through fan communities in the Indian context is that these videos are/were mostly (or used to be mostly) consumed on inexpensive phones (until 2015), and thus the capacity of consumers to modify content or even edit the videos is/was limited. This is a point that S. V. Srinivas also makes while discussing the political mobilization through YouTube videos about a separate state of Telangana from Andhra Pradesh.[18] That said, the easy video-editing features on short-video platforms like TikTok, Moj, and Josh that have emerged since 2018 (TikTok became popular in India in 2018 and Moj and Josh started soon after TikTok's departure from India in mid-2020) have made lip-synching (or dancing) to a song or Shayari (Indian or Pakistani Urdu poetry) or reenacting dialogue from popular Bollywood films much easier, and therefore memes do spread from fan activity.

Like Jenkins, Tony Sampson is also against the idea that linguistic terms be used to relate social and biological phenomena, or in other words, comparing the spread of viral texts with the invasion of microbial species in humans. However, unlike Jenkins, Sampson simply does not comprehend "the social" and "the biological" to be separate: According to him, they are mutually intertwined. Sampson claims that the "monadolgoical" theories of Gabriel Tarde characterize "the social" for the most part, as an "involuntary association, with all manner of affecting agencies."[19] By asking questions like "When we think about what is it that spreads, are we also not considering what we feel about what spreads?" Sampson makes the role of moods, feelings, and emotions key to the spread of viral video and/or affective contagion.[20] Feelings created by and circulated through a viral media text lie at the multisensory intersection point of the social and the biological. Thus, with any viral event, Sampson finds an intermingling of affective social phenomena and hardwired biological responses.[21]

While Sampson's theory of the circulation of affect is somewhat biased in favor of studying the accidental nature of momentous phenomena or chance encounters, I do find his relational framework, which gestures toward affective atmospheres and eschews neat divides of the social and the political, to be both sophisticated and convincing. That said, what sociopolitical structures and contexts have to be in place for the buildup of the affective atmosphere remains underdeveloped in Sampson's work.[22] This is where I believe the affective relation between the state and the citizen and the constitutive role of mediation in the relationship is crucial to study.

Circulatory practices in India are also inflected by "the political," wherein traditional gatekeepers such as the organized mainstream media are met with distrust and there is an affective relation between the state and citizens that has been undergoing significant churning in recent years.[23] Mediation both represents and constitutes the affective relation between the state and the citizen around questions and concerns of corruption, mythical religio-cultural wounds, and the Indian nation's contested past. This chapter is therefore interested in thinking through viral and spreadable media in both the sociobiological and the political and cultural realms.

Sangeet Kumar locates the spread of memetic texts and viral videos on the Indian web related to political satire within a longer historical context of "argumentative culture" in the country, where humor, laughter, parody, and comedic genres have been important for critiquing political and religious beliefs and arguments from within.[24] One could see political parody and satire on television shows along with the presence of stand-up comedians on the Indian web from early 2000s onward. Aswin Punathambekar and Sriram Mohan discuss how the Tanglish (Tamil and English) lyrics of the song "Kolaveri D."—released as part of the publicity for the film *3*—attained popularity throughout India. The first line of the song, "Why this Kolaveri," repeated throughout the song as a refrain became part of numerous mash-ups, remixes, and tweets as politicians, journalists, and citizens used it to make sense of and express their bewilderment, irritation, and exhaustion with political corruption and the general political culture of the country. The song's opening lines, according to Punathambekar and Mohan became a sonic cue, a "sound bridge" to connect the popular and the political.[25] Anirban Baishya writes about the sticky temporalities of memes that mocked the demonetization drive of the BJP government and its media coverage.[26] Some of these meme variants had appeared earlier related to another set of incidents but were repurposed as a response to contingent circumstances and then resurfaced to become critiques of the various absurd theories and (il)logics that were provided to support the demonetization policy.

While short-video platforms are the rage in India (and the world) today, the history of short videos in India has a much longer history, beginning with (analog) VHS tapes and pirate (digital) CD stores in the late 1980s and early 1990s and progressing to Nokia multimedia phones in late 1990s and early 2000s.[27] From *Tehelka* magazine's sting operation tapes documenting political party heads and military officers accepting bribes to MMS (multimedia short message) scandals around pixelated videos of sexually explicit content about school children, the fast-moving and volatile video circulation in India through the decades of 1990–2010 disrupted, as Ravi Sundaram has persuasively argued, the postindependence media order and public culture. Sundaram also notes "that the nervous, playful movement of the pirate aesthetic anticipated meme culture of the post-2010 era."[28] This aspect of history is important to consider because text messaging services (SMS) have been available on cell phones for a while now and have in the past led to riots. However, such messages were not easily affordable, and sending digital images through them is/was cumbersome compared to (the more contemporary forwarding practices on) WhatsApp.[29] Sahana Udupa and Fathima Nizaruddin's research has shone light on how instant messaging apps/platforms have been a key part of circulating extreme speech, misinformation, and violence targeting minority communities in India.[30]

Vigilante Virality on WhatsApp

Since 2014, vigilante actions involving *gau-rakshaks* ("cow protectors" or "cow vigilantes") publicly beating up purported cow slaughterers (also referred to as "cow smugglers") have been on the rise.[31] This is part of a recent upsurge of Hindutva nationalists using alleged violence against cows as a pretext for anti-Muslim and anti-Dalit pogroms. Furthermore, these atrocities are videotaped and made to go viral through WhatsApp. Cow vigilantes tend to be upper-caste Hindu men; they have been extensively covered by mainstream media as being the new "angry young men" who are waging a holy war against cow killers and for whom the cow, the blessed bovine, is a symbol not just of Hinduism but also of India.[32]

While the end-to-end encryption of WhatsApp allows privacy for users, it constricts access to messages. Regulatory problems are compounded further because even the local police and civil society watchdog groups are unable to track objectionable messages unless they themselves are part of the groups in which such messages are being circulated or somebody brings those messages to their attention.[33]

In a Quint documentary, *The Making of Lynchistan: Inside India's Deadly "Gau Raksha" Network*,[34] Rameshwar, a Bajrang Dal district-level convener, confesses to administering 250 WhatsApp groups.[35] Of those, 150 are groups related to cow protection; some go by the name of "Gau Chetna"

(Cow Awakening), "Raj Karega Hindu" (Hindu will rule), and "Antarashtriya Hindu Parishad" (International Hindu Council) as evidenced by figure 3.1, which shows the mobile phone of one such convenor.

Many cow vigilante outfits maintain multiple WhatsApp groups where information about cattle being carried by Muslim truck drivers is spread. Enthused Hindu men are exhorted to converge on particular locations on interstate highways to catch alleged cow smugglers. These WhatsApp groups are considered an integral part of the *gau-raksha* (cow-protection) infrastructure. For example, consider the message spread by the WhatsApp group "Dadri Gau Raksha Dal" (Dadri Cow Protection Party), a vigilante group based in the Dadri village of Uttar Pradesh (see figure 3.2). This image is not of an actual WhatsApp group but was created based on research inputs by the Al-Jazeera Listening Post team as clarified by executive producer Meenakshi Ravi. Here, Dadri youths are being told that a cow has been cut in their village, Dadri, and that to help catch the culprit they should meet at the bus stop at 8 P.M. Furthermore, there is a line at the end, which encourages readers to *phailaye* (spread) the message. The organized aspect of today's cow vigilantism is crucial to consider, because often the violence is blamed on rumors spreading on WhatsApp and thus made out to be something that was instantaneous and viral instead of meticulously planned and executed.

Figure 3.1
WhatsApp groups on a Bajrang Dal member's phone (Quint documentary).

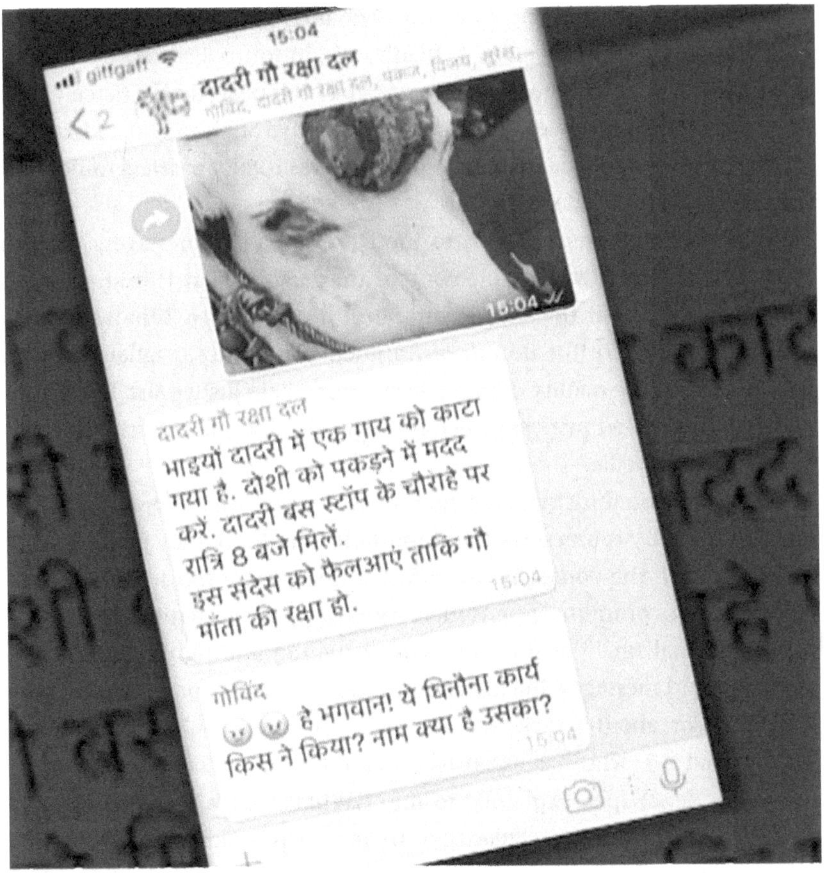

Figure 3.2
WhatsApp message from Dadri Cow Protection Party (snapshot from Al-Jazeera documentary). Courtesy: Al-Jazeera.

This is what happened to Mohammad Akhlaq, who died in Dadri in western Uttar Pradesh in late September 2015. The *afwa* (rumor) that he had been spotted throwing cow-meat leftovers spread widely on WhatsApp. The whole village had pictures of cow meat on their phones that they received through WhatsApp; the meat was alleged to have been found near Akhlaq's house. The pictures were gory and abstract and revealed nothing about whether they had been taken from Akhlaq's house or refrigerator. Soon, according to accounts by two journalists, a mob descended, prompted or provoked by the village temple's loudspeaker announcement that a Muslim man had been found eating cow meat. The mob grew out of nowhere and raided Akhlaq's house; even as Akhlaq's wife, in journalist Betwa Sharma's description, cried, "It is goat, goat, goat," the mob beat Akhlaq fiercely.[36] As Moti Singh, in an interview with journalist Supriya Sharma said, "Khamkha ki afwah thi, iss cheez pe hadsaa ho gaya" (It was totally baseless rumor and the incident happened on it).[37]

WhatsApp management fails to limit the spread of messages, and the Indian government was unable contain the majoritarian Hindu paranoia and insecurity about the supposedly Muslim cow stealer. WhatsApp adds viral indignation to this notion of (im)mediated justice. Vigilante justice has always had the quality of immediacy. Immediate justice also has a component of mediated justice: justice through media or, better still, through instantaneous media.

Such communal incidents of mob behavior are often viewed as "flare-ups" triggered by rumors. Yet it is important to note that for the triggering rumor to have the consequential effect it does, there has to be a regular buildup of the communal atmosphere. WhatsApp is foundational to "everyday communalism,"[38] as it enables cow vigilantes and right-wing trolls to regularly send messages that are communally inflected (for example, about love jihad or about attacks on religious minorities in Islamic nations). Mohammad Ali, who has extensively reported on the Dadri incident for the *Hindu* newspaper, explained to me, "[WhatsApp] is a daily factory of messages, and there is a regularity of them which builds up to something till there is a trigger."[39] This "trigger" often tends to be some local incident that demands an action or reaction. This aspect of organized everyday communalism is key to understanding how today's vigilantism is coordinated.

The Dadri incident and its catalytic buildup is where the Tardean logic of accidental or momentous affective encounters—gathering a crowd of users who are almost entrained to imitate one another in synchronization—seems to both make sense and not make sense. Let me trace both the logics.

The abstract and abject pictures of the meat led to visceral reactions. Socioreligious associations intersected with embodied feelings, resulting in the spread of an affective contagion. The affective atmosphere aspect of the incident is best captured in this quote from Rakhi, a Hindu believer who was petrified at the thought of somebody killing a cow but also admired Akhlaq as a neighbor and his work as an ironsmith: "There is something about the food we eat, the air we breathe today, and all these mobile phones, which makes the young men so frustrated and violent."[40]

And yet, the Dadri incident should not be analyzed as autopoiesis; it required a careful foundation and involved a whole platform ecosystem. This ecosystem consisted of vernacular Hindi newspapers that day in and day out carried stories of cow-meat bans (*gau maans pratiband*) in other Indian state and the village temple's loudspeaker, which pronounced someone guilty without any forensic investigation of the meat. Here again we see that the instantaneous flare-up thesis requires a gradual buildup logic. This gradual buildup logic has also been conceptualized as "fire tending" by Paul Brass in his study of how "fire-tenders" keep the ground perennially ready for communal violence to occur in institutionalized riot networks.[41] Fathima Nizaruddin has identified WhatsApp to be a key platform of fire tending for the communal riots in North East Delhi in 2020.[42]

It is important to emphasize here that the gradual buildup logic should not be dissociated from affect. This is because affect does not always have to be linked to spontaneous, momentous, and accidental events. Rather, affect can be part of daily routines, patterns, and habits. Relating social media habits and affect, Wendy Chun has argued that affect shapes and is shaped by habitual microactions (automatic embodied responses not needing any reflection) such as tweeting, liking, updating, and sharing.[43] WhatsApp forwards are a socially habituated way of doing things with mobile phones, and a lot of rural Indian youths participate. Affects are drivers of habits, and WhatsApp forwarding becomes a habitual microaction. These daily WhatsApp forwards thus become a key part of techno-political affect and should be considered as much a part of the affective theater of cow vigilantism as the more jolting, disturbing, and momentous encounters portrayed in cow vigilante videos.

The more everyday WhatsApp forwards may not even be negative messages about the Muslim community but just good morning messages extolling the virtues of "mother cow" (*Gau Mata*). As Radhika Govindrajan notes, there is one group of militant *gau-rakshaks* (cow protectors) and another group of *gau-sevaks* who do "seva" (service) toward cows. For some, this cow service could take the form of attending to the daily needs of a cow and volunteering time at a cow shelter, while for others, this could be just about sending images of cows with feelings of *bhakti* (devotion). This quotidian microhabit of forwarding messages about cows also builds toward coalescing a community that may not take to the streets in the form of mobs attacking minorities but is sympathetic to the efforts to protect cows.[44]

I have explained in this section that discerning the WhatsApp networking effects is crucial to understanding how cow vigilantism is premised on both momentary/accidental affective encounters and on the daily and accumulative affect of habitual WhatsApp forwarding practices. There is no one reason for the unabated WhatsApp forwarding tendencies but, rather, a conflation of various entangled reasons: habit, pressure to participate, prejudice, and (in)adequate media literacy.[45]

This senseless prejudice against minority communities and the zealousness with which such mob lynchings are carried out, which are then unabashedly recorded and even used for recruitment drives, can be understood as instances of aspirational hatred. Arjun Appadurai conceptualizes contemporary India's aspirational politics through the term "aspirational hatred," where jobless youths are seduced by dreams of national, ethnic, and religious purity to express hate toward those to whom they feel superior.[46] The youths see their own community's religious and caste leaders spreading hate, feeling proud, and enjoying impunity, so they aspire to imitate their leaders so as to gain recognition in their community.[47] This also can be seen as an instance of what Priya Chacko calls the "populist politics of resentful aspiration," where lower-middle-class young Hindu men are supposed to feel resentment against Muslim men, whom the previous established elites belonging to political parties other than the BJP had supposedly pandered to.[48] Youth dissatisfaction is appropriated for stoking resentful aspiration and thereby inciting violence against minorities.

The virality of messages also has to do with certain affordances of the WhatsApp platform: the potential broadcast size of groups, administrators' ability to add more people to the group, and the tendency of group members to keep forwarding messages. Acknowledging this, WhatsApp has taken some steps to limit the number of forwards to five (in India), and there is an option (though not the default option) provided to WhatsApp users to not be added to groups without their consent.[49] Digital rights activist Nikhil Pahwa suggested that WhatsApp should consider giving the original sender of the message the power to decide whether it should be forwarded.[50]

Many of the videos circulated on WhatsApp also travel through YouTube, and YouTube over time has acted as a repository/archive of those videos. There are videos posted by cow-vigilante outfits on YouTube/Facebook (or forwarded on WhatsApp) that show them brandishing weapons and chasing trucks suspected of carrying cattle, including that of fifty-five-year-old dairy farmer Pehlu Khan, who was killed in Alwar's Behror area in the state of Rajasthan by one such *gau-rakshak* group. Gowhar Faroq has written about another kind of stylized performative spectacle in the form of YouTube music videos made by microcelebrities and so-called creative entrepreneurs.[51] These music videos demonize Muslims with lyrics full of expletives, create fictionalized chases of purported cow smugglers, and try to enlist viewers to volunteer in efforts at protecting cows. Kunal Purohit has investigated and documented a whole ecosystem of songwriters, singers, studios, and production houses that produce Hindutva-fueled entertainment and propaganda in the form of songs and music videos.[52] Other mass entertainment and popular cultural forms such as poetry and book publishing are also identified by Purohit as "Hindutva Pop" that provides an "unending supply of everyday communalism" that slowly builds up to the spectacular violent events.[53]

From 2014 onward, while the cow vigilante videos were mostly a YouTube phenomenon, many videos of the Delhi riots of 2020 cropped up on TikTok. Several of these videos used TikTok's "sound" icon or the "Use the Sound" feature for imitation and replication, where lyrics of songs from one context were problematically used to frame protests and riots in a different context. Before discussing the specifics of this, I want to begin with some features of the TikTok app and its reception in India. After all, WhatsApp was not just used for vigilante virality, but served practical benefits of affordably sharing vernacular music videos, talking and chatting among friends and relatives, interacting with business clientele through WhatsApp Groups especially for rural women entrepreneurs during Covid-19 pandemic, and keeping in touch with family across transnational scales.[54] Similarly, TikTok should also not be identified as merely a problematic meme factory but be recognized for the tools and venues it offered for expression to millions of ordinary Indians across caste and gender barriers who (for the first time) were able to showcase their talents and present their lives. The easy imitative aspects of TikTok's challenges feature were also leveraged by brands and marketeers for promotion.

TikTok Memes

If forwarding is the key way for spreading, sharing, and socializing on WhatsApp, TikTok offers many different ways of spreading content, especially through video creation that involves miming or parodying other videos, often set as challenges. Years of accumulated (and continuing) Bollywood film culture acted as an archive, or in more capitalistic/opportunistic terms, a resource, to mine and mime for TikTok-ers who put their digital bodies at play on the mobile phone screen. Whether it is a particular hook step of an iconic Bollywood film's song-and-dance sequence, a Bollywood superstar's popular signature gesture, or a specific style of delivering a punchy one-liner, each of these cinematic moments lend themselves well to reenactment and imitation on TikTok to become postcinematic digital fragments.[55] In India, with Bollywood celebrities participating in dance or lip-synching TikTok challenges, their fans and other social media microcelebrities and influencers followed up with imitation and reaction videos, increasing their visibility among ordinary Indians. Themes, brands, and videos were spread and networks were formed through what Diana Zulli and David James Zulli call "imitation publics," actively supported by the digital structure of TikTok, which helped initiate such imitations.[56] This viral strategy became another way of campaigning for advertisers beyond the usual channels of television and newspapers.

In April 2019, as summer was approaching and temperatures were beginning to soar in India, Pepsi launched an extended TV commercial with a new catchy tagline, "Har ghoonth mein swag hain" ("Every gulp has swag in it"), featuring two Bollywood stars, Disha Patani and Tiger Shroff, who are known for being remarkable dancers, and the singer, rapper, and lyric writer Badshah, all three of whom are very popular among the Gen Z audience. The long TV commercial had its own narrative, with Patani and her friend trying to enter a party without passes and being stopped by a guard on the door. That is when she finds Tiger Shroff outside on a bench with a Pepsi bottle; he says that for a party one only needs swag and throws the bottle to Disha, implying that there is swag in Pepsi. Disha queries, "Aur music?" asking about where the music is, to which Badshah, who seems to have been operating a makeshift food stall, suddenly reveals DJ equipment and starts the music. An elaborate song-and-dance routine follows on the beats of "Har Ghoonth Mein Swag Hain," penned and sung by Badshah.[57]

Pepsi started the TikTok campaign a month after the release of the initial long-form television commercial. In May, the "Har Ghoonth Mein Swag" campaign in the form of the #SwagStep challenge was kicked off by the stars of the music video, Disha Patani and Tiger Shroff, who released selected short versions consisting of dance moves from the original TV commercial sequence, including the signature hook step. These videos were then shared by their followers on TikTok and led to fifteen thousand user-generated videos in just one day.[58] Toward the end of many of these replicated-with-some-variation dance videos, the influencers and microcelebrities also took a gulp from a Pepsi bottle, thereby endorsing the brand (see figure 3.3).[59]

By mid-2019, advertisers in India had learned how to run campaigns successfully on a short-video platform like TikTok with a group of brand ambassadors that included film celebrities as well as TikTok influencers and microcelebrities. For a while, some advertisers shied away from TikTok, regarding the app with some hesitation because of perceptions about the lack of content moderation. However, that seemed to have changed, as *The Economic Times*, the most influential business newspaper in India, published

Figure 3.3
Pepsi's "Har Ghoonth Mein Swag" campaign challenge on TikTok by Disha Patani and Tiger Shroff.

Figure 3.3
(continued)

an article titled "On TikTok, Everyone Is Up for a Challenge" in August of 2019. In the article, experts opined that brands had found TikTok challenges to be a technique of indirect social persuasion where "influencable" people feel called upon to try out a dance step, with the incentive of being able to prove that they can also do it or they can do it even better.[60]

For a while during its course of diffusion in India, first as MusicAlly and then as TikTok, the app remained under the radar (of the elite mainstream) as it found its initial audiences and users in provincial India among those who belonged to the marginalized castes and classes, and who for the longest time had remained invisible on the Indian internet.[61] Many have praised TikTok for providing a place for "lower-caste" and rural female influencers in India to express themselves and receive online recognition. However, these new social media stars have been trolled by bigoted "expert" commentators such as popular YouTube star "CarryMinati," who regularly joked about the appearance of—and the clothes worn by—the emerging content creators of TikTok. Other so-called elites on Instagram and YouTube mocked the aesthetic style of some of the videos on TikTok and dubbed them as crass or cringe.[62]

Some aspects of the backlash to these interclass and interregional neomobile users and their uploaded content on video-sharing apps has also been witnessed in China with its own historical and contextual specificities. In China, more than TikTok (Douyin) (which caters to middle-class sensitivities in that country), it is the Kuaishou video-sharing and livestreaming app that has attracted rural youths in tier 3 cities to engage in cultural production and showcase their "authentic" village settings and lives. These videos have caused moral panics, especially among upper-middle-class Weibo (microblog platform) users who label the Kuaishou productions as *tuwei*, meaning tasteless, unintelligible, and absurd. At the same time, *tuwei* could mean "earthiness," suggesting a grounded connection with the rural, and these video-makers and live streamers (*zhubos*) have fans who experience a sense of cultural intimacy while watching these productions.[63]

TikTok's mimetic infrastructure can be appropriated for spreading hate, especially the response videos and duet features. In the town of Thalavedu, located in the state of Tamil Nadu, caste hatred spread through TikTok response videos. Leaders of particular caste groups such as the Nadars and Vanniyars made videos ridiculing Dalits and asserting their dominance over Dalits. Dalits are considered lower-caste in the entrenched caste ladder in India's deeply unequal society. Dalit community members in Thalavedu responded to the hateful videos about their community by creating duet response videos directed against the Nadars and Vanniyars. Nilesh Christopher reported about a Vanniyar community member, who responded to hate against his community by creating a fifteen-second video that said, "You Paraiyar dogs stab people in the back. Our courageous lot meet people face to face. I'll chase each one of you Paraiyar down and slay you."[64] This threat of slaying is made to "Paraiyar dogs," a humiliating reference to Dalits. Dalit denizens of the town that Christopher spoke to noted, "We lip sync to songs and put up funny speeches of politicians, but if we see a video dissing our community, we respond with a duet abusing them. We can't be on the receiving side forever."[65] A particularly common video trope Christopher observed was users making a "duet" with a fifteen-second clip of urinating into a toilet next to videos of popular political icons of rival caste communities. Dalit civil rights icon Bhimrao Ambedkar was also insulted by such problematic duet videos, where his image was presented next to footwear. Practices of tanning and making leather shoes have traditionally been forcibly consigned to Dalit workers, so this duet video was supposed to be an insult.

Therefore, what we observe here is that "imitation publics" formed through TikTok can help an ad campaign go viral by making a hook step trend; at the same time, a different formation of "imitation publics" can become echo-chambers of hate and ridicule as responses and counterresponses to casteist threats pile up and escalate into physical fights. Indeed, death threats issued on TikTok as part of a response to competitive chest-thumping by rival caste group members resulted in fights, murders and deaths in Thalavedu.[66] During a conversation with me, Nilesh Christopher elaborated on his article on TikTok and caste assertion in Thalavedu. He explained that Dalits in Thalavedu are a proud "congregation of Ambedkarites" who want to "own the word Dalit" and like lip-synching to songs in praise of Ambedkar.[67] In one sense therefore, TikTok helps in sharing Dalit songs and assists caste members to create a cohesive community identity. However, when these same songs enter the local-level TikTok feed of other caste members in the same town, they become a problem. These lip-synched expressions of Dalits irk the Vanniyars, who see such Dalit caste assertion as a challenge to their dominance. Beyond the binaries of viral advertising or fueling an epidemic of hate speech, TikTok has a nuanced history of user engagements in India.

Journalist Shadma Shaikh ascertained that much before Chinese short video platform and livestreaming companies had set foot in India in 2016, there was already a consolidated and thriving industry of such platform companies in China, with talent agents and content creators.[68] With the benefit of such industrial experience in China, TikTok and BigoLive (Singapore) endeavored to attract talent to create content or host sessions for their platforms in India. Indian content creators performing in these untested platforms, with a combination of eagerness and reluctance, were getting used to being paid based on contracts or performance targets. Performance targets included being able to post a certain number of videos per month, and the payments depended on the number of likes those videos received.

On live streaming platforms, target-based payment criteria included hosting shows for a certain number of hours and days per month as well as the number of people joining the live streamed shows.[69]

App entrepreneurs in China told journalist Shadma Shaikh that something similar to what was happening in India after Jio's 2016 launch had happened in China: When internet prices dropped in 2010–2011, there was an increase in content apps. Based on their experience of catering to new smartphone users, a Chinese app professional told Shaikh that neomobiles tend to prioritize "entertainment" over "communication": "When users come on the Internet for the first time, what they are looking for is entertainment. Of course, they are looking for to communicate with people, but that is something that a feature phone might also offer them. When they have Internet, the first thing they want is entertainment."[70]

Therefore, it is telling that on TikTok, platformed sociality is organized less around interpersonal connections and more around entertaining content. To put it differently, "creative interaction" (that is, interaction related to commenting on or imitating a video feature style, edit, or filter) takes precedence over "discursive interaction" (that is, platform users are not being encouraged to deliberate on a topic) in TikTok's platformed infrastructure.[71] Homegrown short-video platforms in India, like Moj, also follow TikTok's lead in designing their algorithms, which value content and creative interaction over personal connections and discursive interaction. Moj's head, Ankush Sachdeva, keeps emphasizing that like TikTok, Moj users are shown recommended content based on their preferences, viewing history, and real-time usage (feed-ranking algorithm), which differs from recommendation algorithms used in Facebook and Twitter, where a majority of the content that individual users see is based on who they follow or are friends with.[72]

Moj and Josh: Short-Video Platform Futures

The Moj app's parent company is Mohalla Tech, which started Sharechat, the first regional user-generated content-sharing-based social media platform in India. For a long time, social media entertainment in India was restricted to English-speaking users. Facebook and Instagram were tailored to English-speaking urban Indians, and not enough platforms catered to the regionally speaking audience, except for YouTube, which to some extent reaches across tier 1, tier 2, and tier 3 cities.

Among homegrown Indian apps, it was ShareChat, launched in 2015, that first created a social network for India's regional language users and heavily targeted non-English-speaking neomobile users who found the English language internet to be "unfriendly" and/or "unfamiliar."[73] ShareChat had simple templates for creating content that could be easily shared even on WhatsApp and Facebook, which meant memes in regional languages could circulate.[74] For a long time, venture capital in India was reserved for urban elitist platforms, and ShareChat definitively changed that. Several tech journalists who had earlier reported on ShareChat said that when they were asked to write about the app and looked at the content, they "thought it was so lame because we [they] could not relate to it." One of them noted that they wondered, "How is this [Sharechat] getting so much money or even attention?"[75] So elite urban users who used Facebook and Instagram could not relate to the ShareChat content, but this was the beginning of the time when new (mobile) internet users were coming in, and there were no other platforms that were catering to them. When TikTok left following the Indian government's ban in June 2020, it was no surprise that ShareChat, given its experience in the regional social media space, decided to create one of the TikTok clone short-video platforms, called Moj.

Many venture capitalists, app developers, and tech journalists I met bemoaned the TikTok ban in India, noting that the ban had caused irreparable damage to the rise of regional influencers whose creative aspirations TikTok had nurtured in small towns in India. There were some who maintained that there could not possibly be another app that could replicate TikTok's success and some who were less skeptical and believed that in two to three years, the TikTok clones would be able to recreate that kind of success. To match TikTok's success, these new short-video platforms had to make investments in AI technology as well as up the ante regarding on-the-ground talent recruitment. What has happened in the years since TikTok left is what I will discuss in the last section of this chapter, along with the implications of such developments for shifts in scales of sociality and spreadability as well as influencer marketing and monetizing. Has Moj, or other short-video platforms like Josh, been able to come close to the success TikTok had in India? In particular, I focus on these companies' talent acquisition efforts and how they impact platformed socialities, aspirational and entrepreneurial labor, and the distribution of content.

Raghuvendra Prajapati, who Moj hired to help with content acquisition, explained to me the relationship between categorizing audiences (into the tier 1 cities, tier 2 small cities, tier 3 towns, and villages) and hiring content creators. He mentioned that knowing your audience is helpful because, depending on the type of audience, the platform will serve content tailored toward them, and this impacts the kind of content creators the platform will hire: "Isliye categorization kiya jata. Kis tereke ka audience. Content ka jo flow dena, kis type ke content creators ko hire karna hain" (That is why audience categorization is done. The kind of flow you need to give to the content, the type of content creators you need to hire).[76] Prajapati described the audience segmentation in terms of taste cultures, saying that if a tier 1 city audience member in Bangalore is listening to the South Korean boy band BTS, by the time you reach a tier 3 city (more of a town) and ask the person about BTS, he will say he knows the Bhojpuri singer Khesari Lal Yadav but has no idea who this BTS is. Like their audiences, content creators in tier 1 areas know about BTS, and those making content in tier 3 towns are dancing to and mimicking Khesari Lal's music videos. Audience segmentation and localization of content creators is therefore about familiarity and relatability of content. Talent scouts such as Prajapati seemed to occupy a "mid-level managerial role," involved as they were in "managing talent, audience, and genre," thereby bridging "managerial and creative roles."[77]

These audience segmentations predicated on regions and types of phone had an explicit class dimension, but what remained absent were caste-based categorizations, which struck me as a strange omission given the all-pervasive presence of caste in everyday social life in India. I put this question to Prajapati and have asked other content creators and content heads about this as well. The answers I have received vary: Some said caste-based targeting of users is something politicians do for votes and is not a practice encouraged in the corporate industries of India. Some suggested that because short-video app algorithms for targeting users were imported from Western countries where there is no concept of caste, these algorithms did not provide for caste-based outreach. Some accepted, though, that at the level of individual creators and their fans, there certainly was a caste dimension: a microinfluencer of a particular caste was likely to have more followers of that caste. However, they refused to generalize about caste segmentations, deeming them to be too traditional or too antithetical to the values of Indian corporate start-up cultures.

Prajapati still does some content creation for Moj, but his primary job is talent acquisition and content direction. His creator handle on Moj is "Kanpurwala Raghav," a way of gesturing to his local authenticity as hailing from the city of Kanpur. Some of his content also applies the colloquial idiom of the Hindi dialect spoken in Kanpur (see figure 3.4). Prajapati used to be a content creator for Roposo, an Indian social video-sharing app, but was not earning much from being in that app. His father had lost his job during the pandemic and, in need of money, he was aiming to go to Bangalore and start working for the food delivery company Swiggy. Around that

Figure 3.4
Raghav's Prajapati's content handle at Moj platform as Kanpurwala Raghav. Courtesy: Raghav Prajapati.

time, Moj was looking for people to join their creative team to help with both content direction and creative talent acquisition.

Something about Prajapati's resume impressed the Moj team: his background in theater. Having been part of a theater troupe in college, Prajapati had traveled around India by train, visiting many places and performing for audiences in various states. Prajapati's theater contacts were helpful in

Figure 3.4
(continued)

discovering new content creators. Moj provides a lot of incentive to travel, and Prajapati's role as a talent scout requires such travel. Prajapati's work from home desk job is from Monday to Thursday, and the Friday to Sunday breaks provide him a long weekend to travel. He has encountered talented content creators walking in the streets and traveling in trains, which is something he documents at times as well (see figure 3.5). Chatting on trains, if he finds somebody narrating a compelling story or uttering a comedic punchline, he asks for their phone number and shares it with his creative team at Moj, who then onboard the new discovery as a content creator after a couple of rounds of interviews.[78]

Prajapati deploys the Hindi/Urdu word *atrangi* (connoting something "unusual") often while discussing the exercise of spotting talent, especially those who seem to be doing something out of the ordinary, such as juggling balls. He asks individuals practicing such unusual activity to consider being a content creator . Prajapati explained to me the way he describes the sequential steps toward professionalization and entrepreneurship to the newly initiated or onboarded content creator: the way to gain followers, the prospects of earning an income from content creation, the ways to build relationship with brands:

> Ek set pattern hota, jaise hum padhai kartein hain uska bhi ek pattern hota hain. Waise hi content creation ka ek pattern hota hain. Kaise aap follower gain kar sakte hon, kaise then income create kar sakte hon, kaise brands aap ki taraf aa sakte hain. Toh yeh jab cheezein thodi si explain kar dete hon, halka sa samnewale ko monetary benefits dikhta hain . . . thoda sab hi monetary benefits . . . arre haan bhai, yeh toh hum aise hi kar rahein hain . . . toh kyun na hum iss ko iss format mein bhi kar dekh lete hain . . . phone sabke pass mein hain. (There is a set pattern. Just like when we study, when we create content, there is a pattern. How you gain followers, and then how you create income, how then brands approach you? So, these things when you explain these things, and then the person in front of you can see the monetary benefits . . . so they go, why not try this format, everybody has phone.")[79]

Prajapati's talent scout work involves being a cultural intermediary as he assesses and explains, or still better translates across, the commercial and the aesthetic value of particular created content. Kyle Barnett has written similarly about the managerial intermediating role played by talent scouts for phonograph recording companies in the United States between the end of World War I and the start of Great Depression. These talent scouts "moved through wildly different segments of society" and identified

Figure 3.5
Prajapati traveling in search of content creators. Courtesy: Raghav Prajapati.

Figure 3.5
(continued)

and brought new genres (now called jazz, blues, and country) to the record labels and recruited new talents for them.[80]

These different creator acquisition efforts are not unique to Moj, but other short video platforms such as Josh and Chingari are also working on making themselves an accessible destination for talented content creators.[81] Seher Bedi, who heads the Josh Studio also elaborated on the various initiatives that Josh was taking toward nurturing and scaling talent from the hinterlands of India.

Bedi talked at length about sending mobile stages set on canter vehicles to the deep interiors of India for their creator recruitment drives as part of the Josh Studio's initiative called "World Famous." The following words were inscribed on the canter vehicle: "Talent Dikhao, Famous Ho Jao" (Show Talent, Become Famous) (see figure 3.6). "World Famous" is a talent hunt that involves the mobile stage on the canter traveling to many small towns across states such as Gujrat or Rajasthan.

> So, we have a property called "World Famous" . . . and what World Famous does is . . . you know what we are going to create the next 1 million stars of Bharat, and what that is . . . we go state by state, so we take a Gujarat, and we say while we have the finals and semi-finals in the big cities we look at the hinterland . . . we look at over 70 locations per state in the hinterland regions . . . these little towns . . . where we send in a canter . . . and canter is just like a mobile stage which people can just jump onto and showcase their talent.[82]

Figure 3.6
Josh mobile canter vehicles "World Famous" for talent hunts. Courtesy: Seher Bedi.

These mobile stages became the venue for local people from tier 3 cities to perform and showcase their talent. Bedi excitedly mentioned discovering such dynamic youthful talent and so many different new performance genres (beyond comedy, music, drama) and modes of expression (including *shayari*—particular forms of reciting poetry) that she learned about from this initiative. She declared that she would not have known about these talents and new genres of expression had she confined herself to Mumbai. The emphasis "on creating 1 million stars of Bharat" is crucial as a way to differentiate "Bharat" from urban elite "India" which goes well with the rhetoric and action of outreach (away from Mumbai) toward the hinterlands, where the essence and earthiness of Bharat resides.

While earlier initiatives to tap talent for reality TV shows had experimented with renting out space in shopping malls or cinema halls in tier 1 and tier 2 cities, it still restricted the outreach efforts to tier 3 cities where there were no halls and malls. The mobile stage on the canter proved to be a more successful way of conducting auditions ensuring that both talented content creators and their admirers came to the mobile stage traveling through their village to perform: Bedi mentioned garnering a "footfall of 1–2 lakh" people in each destination the canter stopped. Bedi seemed excited about not only acquiring talent in this manner but also being able to scale Josh's content so that people in many different parts of India can see their view of what their world is.[83]

Both Josh and Moj create retreats and workshops where talented creators can meet other creators and influencers, and in many cases even mentors, fostering collaboration. Josh's vertical video short-film festival JFlix asked for one-minute film entries as a way to challenge long-narrative storytelling practices. The finale, which involved a red carpet award ceremony, featured Bollywood superstars who gave awards to the Josh short-video film winners in various categories. The short-film festival was also advertised as an opportunity for content creators and influencers to collaborate with mentors. Bollywood directors provided a master class and exchanged ideas with content creators.[84]

These gatherings can also be seen as a space where Josh's content creators can get to finally meet Bollywood celebrities. These events certainly have their precedent in reality TV show contests in terms of the talent hunt procedure followed and the much-anticipated meeting with a celebrity, but there is a faster pace to this platformed sociality: an immediacy to create and edit content and then post it on Josh and then be selected to receive an award and receive recognition from and converse in-person with a Bollywood celebrity.

At an earlier moment, television as a technology of media saturation fostered a way for a fan and a celebrity to have an imagined closeness or for the fan to entertain the illusion of unfettered access, often referred to as a parasocial relationship. Alice Marwick has remarked that social media has amplified such affective relationships, where the "parasocial" has been transformed into the "potentially social" with accelerated emotional ties between celebrities and fans.[85] Short-video platforms have allowed their ordinary app users to think they have a potential social relationship with microcelebrities, and these microcelebrities in turn have a potential social relationship with the Bollywood celebrities.

A content director for one of the short-video platforms once told me that people in India desire a kind of closeness or proximity to celebrities, which they believe also firms up their identity and makes them famous among their neighbors, friends, relatives, and followers. In an earlier moment, before the advent of short-video platforms, according to this content director, to become famous in one's village or town among one's neighbors and friend circle, a person had to come to Mumbai, start looking for work in films, and while walking on sets, come across a celebrity and then take a selfie with them to send home to friends. The content director noted that short-video platforms have made this process much easier than before: "Yeh short video ne iss cheez ko itna easy kar diya hain . . . ki inke through Mumbai mein audition liya ja raha hain—logon se bola jata hain ki video bejh do—chote chote gaon ke log viral huin" (This short video has made this thing easy . . . through them there can be an audition in Mumbai . . . people are told to send videos . . . people from small small villages have gone viral).[86]

During our conversation, I mentioned to Seher Bedi that it seems that content creators are (or can be) fans wanting both to have their own fan following on the short-video platform and to reach out to and be recognized by the celebrities whose fans they are. Bedi agreed and emphasized in her reply the aspirational dimension of the facilitated encounter between the content creator fan and the celebrity they look up to through events such as the creatorthon organized by Josh. She recounted one such aspirational story of a rapper, Vijeeta, from the small town of Karnal in the state of Haryana, who could not believe that he would ever get a chance to work with one of India's most famous rap artists, Badshah. This meeting of Vijeeta and Badshah was made possible because of Josh's creatorthon, an initiative pairing content creators with their mentors or gurus for a collaboration:

> "Definitely aspirational, I think by doing initiatives like you know creatorthon we had Badshah and we got you know . . . the type of people we got musicians from our platform . . . and one of the guys is a rapper, his name is Vijeeta. And he is somebody we got literally from the *gullies* (alleyways) of Karnal. Picked up and you know he meets Badshah who is his biggest *guru* (mentor). He starts creating content with Badshah. The next thing you know is you have got the next big rapper on the scene. That is how quick things turn around. So, aspirational of course—because they get to meet people in their dreams they don't think they are going to meet, and then how their career sort of takes a leap because of that sort of collab, makes it even more thrilling and exciting for them."[87]

Beyond these collaborative events and retreats, talent managers and content directors working in creative teams with content creators exchange notes on a more regular basis over online meetings about what kind of content is trending at the moment. They brainstorm ideas about how to make content that is at once localized, situated, and relatable to platform audiences/users and at the same time serves aspirational visions for both the audiences and the creator. Rosie Findlay and Brooke Duffy have written about this delicate balance between creating relatable authentic content and the aspirational dimensions of creative labor in other contexts in terms of "aspirational realness."[88]

Limits and Aspirations

In this chapter, I have navigated several methodological approaches to understand how the affordances of various platforms, in fostering sociality and connectivity, can make user-generated content move, scale, and spread. In examining the specificity of WhatsApp forwards and TikTok memes, I have sometimes espoused a "platform-sensitive" approach that helps me think of the specific affordances of each platform (that is, WhatsApp or TikTok) in spreading content, and at the same time acknowledges that these "affordances of platforms are relational and multi-layered."[89] Platform-sensitive approaches can become platform-centric as well, and there is a need at times to understand how users perceive and approach platforms, as Mehta and Kaye note in their essay on digital content creation in India.[90] However, content creators are not the only users of any platform, because advertisers, influencer marketers, talent agents, production managers, and content directors also engage with platforms at different levels. While discussing the efforts of short-video platforms like Moj and Josh to replicate the early success of TikTok among new-to-mobile-phone users/audiences, I have tried to balance platform-centric and creator-centric approaches by bringing in the perspectives of talent scouts, content directors, and production managers who act as key intermediaries between the content creators and platforms and shape the trajectory of mobile media distribution.

The field of media distribution studies has often shied away from affect theory in explaining the spreadability of content, but Joshua Braun has noted that it might be "potentially rewarding" to see how the "circulation of affect" and "other forms of circulation might be connected to media distribution."[91] In this chapter, I have drawn from scholarship in affect studies to explore the circulatory practices of instant messaging and short-video platforms. With that said, I have stressed the need to account for the political structures and cultural contexts that exist for the buildup of affective atmospheres that end up influencing platform users' particular micro-habits and usage practices. Doing affect studies does not mean I eschew the political economy of distribution; rather, a media industries approach has been important for this chapter as I have tried to understand how various short-video platforms in the context of India's app ecosystem have had competitive, collaborative, and imitative relationships with one another in their plans, strategies, and efforts to target the emerging neomobile users.

As I try to unpack a particular variant of "mobilities (un)limited" in this chapter—that is, the unfettered circulation of fake news and misinformation online through WhatsApp groups—the readers will have noticed generative tensions between several approaches that I gather together to understand this rather complicated phenomenon. Some observers blame the virality of WhatsApp's cow vigilante videos on regressive Hindutva forces. Others explain such virality in terms of affective circulation through the socio-technical assemblages of the WhatsApp infrastructure. Still another view suggests that WhatsApp use does not necessarily cause momentous mob-like behavior by itself but, rather, that WhatsApp use is habitual and ordinary; through their everydayness, WhatsApp forwards steadily build up an atmosphere (*mahaul*) of rumor (*afwa*) and suspicion, and it is in such a vicious environment that a small communal trigger can lead to extreme violence.

I trace connections across these approaches to explain how global capitalism and right-wing nationalism converge in the steady rise of techno-populism around mobile phones and apps such as Facebook and WhatsApp.[92] Inherent in the mobility of the viral video made by the Hindutva groups is the possibility that it can be intercepted by people of different ideological leanings. The aspirational mobilities of high-caste Hindu men perceived to be achieved by the spread of the video reach their limit when the video appears on the phone of progressive anticaste and secular activists, who then report it to police authorities. WhatsApp's forwarding groups do not allow for controlled distribution, and hence here again, like in earlier chapters, although in a different context, we see the limits of aspirational mobilities. Memes on TikTok and other short-video platforms like Moj and Josh also seem to be marked by a sense of limitlessness in the way they can circulate and scale up very quickly. At the same time, as quickly as memes can trend and go viral, they can also cool down and cease to matter after a few days, until at times, they are picked up again later, when some political corruption scandal or a hook musical tune or dance step is in the news or in fashion again.

Competitive caste and religious politics also lead to the spread of hate memes and problematic response videos in TikTok, utilizing the short-video platform's duet features that mock another caste/religion's leaders and that lead to communal clashes. With that said, virality on TikTok is not restricted to hate and rumors; it also has been a platform associated with viral advertising, reaching more and more people, which is where we see a different kind of aspirational politics, not just aspirational hatred. The aspirational labor of digital creators is fueled not only by sharing their content and receiving the appreciation of their followers, viewers, and subscribers but also by hopes of monetizing their content and someday working with important brands to earn money by participating in their advertising campaigns. Of course, follower counts matter for receiving discounts and freebies or brokering brand deals or gaining prestige. What also matters are follower counts in particular platforms where advertisers are interested in promoting their brand.

Staying in touch with certain content creators over the years now, I have noticed some making the successful transition from being short video platform performers to developing long-form content for YouTube to acting in Bollywood films or OTT TV shows. Not every influencer is able to charm their way into increasing follower counts, with some experiencing downward mobility: They are forced to leave content creation and take jobs in food delivery, and for some even the gig economy becomes unsustainable, and they go back to find work in their uncle's garage or become a night guard in an apartment complex.

Aspiration can be about many things and need not always involve a desire for socioeconomic mobility or a craving for the lifestyle of the upwardly middle class. Some journalistic accounts suggest that a group of working-class creators and influencers who had found a venue for sharing earthy depictions of village life as part of their short video performances on TikTok could not successfully migrate with their followers (post–TikTok ban) to Instagram. InstaReel's curated "bland" images of "aspirational" middle-class lifestyles were too alienating for these content creators, making them feel as if they did not belong there.[93]

Since Yashraj Sharma wrote this in October 2021, it seems that InstaReels has reached out to tier 2 and tier 3 cities. Content creators in these towns have also emerged with innovative tactics of posting across platforms.[94] While maintaining a follower base in short-video platforms like Josh and Moj, these creators cross-post to InstaReels so as to be tapped/spotted by key brands for promotion deals. Shudeep Majumdar, an influencer marketing executive at Zefmo told me that some of the homegrown short-video platforms inspired by TikTok have prioritized localization, but they are no match for Instagram and YouTube: "We don't run any campaigns in these homegrown platforms [Moj and Josh] . . . we do not get any client-request in terms of user-base, that is also questionable those who are in well-established platforms [Instagram and YouTube] . . . de-risk . . . probably want to sign-up and repost the content."[95]

Majumdar claims that influencer marketing does not operate in a significant way for Moj and Josh. He agrees that content creators who are paid by Moj and Josh produce content for them as part of their contract. They then also repost the same videos created for Moj and Josh in InstaReels and YouTube with the expectation that such content will get monetized and gain the attention of key brands. According to Majumdar, these creators by reposting are thereby able to reduce the risk of their content not being seen by sponsors and marketers. In Majumdar's estimation, homegrown short video platforms are nowhere close to the kind of popularity level TikTok scaled to before its eventual ban in India. Even as the owners of Moj claim that they have invested extensively in technology, including AI-related machine learning tools, Majumdar doubts the efficacy of Moj's algorithms in learning about its users. Majumdar's views are confirmed by others who are pessimistic about the chances of Moj, Josh, and Chingari being able to fill the void left by TikTok. These short video platforms have not been able to retain users, and they have had to downsize their workforce because of inadequate revenues and VC funding cuts.[96] Some Moj and Josh officials I spoke with stressed that other ways of monetizing were successful, and that included livestreaming opportunities where microinfluencers and their fans/followers could meet. This venue was where product placements (of advertisers) happened and where fans contributed small monetary amounts as gifts to these microcelebrities. This arrangement allowed microinfluencers to earn money on Moj and Josh and also afforded a whole group of neo-mobile users to participate and feel included through a shift from parasocial to potentially social interactions.[97]

Virality, or spreadability, is as much about the content that goes viral as it is about the practices of spreading afforded by social media platforms. Whether it is WhatsApp forwards or TikTok memes, the content that goes viral often is short videos, and here the scale of the content itself seems to have micro dimensions. That is, the content that circulates at a macroscale is in itself micro, which Jamie Coates sums up by saying, "Scalable sociality and scalable content are co-constitutive."[98] While in the last chapter we discussed streaming long-form content, this chapter has focused mostly on user-generated short video content. Acknowledging the attention spans of contemporary audiences and seeing the virality and popularity of short videos, even long-form content streamers often repackage the long-form content into short videos so that a particular instance of narrative conflict in a TV series or a spicy bite of discussion in a talk show can go viral. Such short videos can serve as a trailer or teaser to draw more audiences to the show.

I began this chapter with tech journalist Pankaj Mishra predicting that the 2024 elections would have been TikTok elections. That did not exactly happen, because TikTok was banned in India, but Mishra's speculation that short video platforms would be a key venue for political message circulation indeed proved to be accurate. Even before the main national 2024 elections, microinfluencers on short video platforms, including InstaReel and YouTube, started playing an important role in political campaigns related to various state-level elections because of their grounded connection with their fans in a particular locality.[99] Microinfluencers in "Bharat" India have a way of using local language and regional markers, and their followers find their political messages to be more relatable. While some political misinformation spread through WhatsApp (via forwards) and TikTok (through memetic challenge and response videos) in India has contained negative and hateful messages, the microinfluencer-inflected political branding in the short video platform space tends to often be positive and understated, sometimes even subliminal. To avoid backlash, some microinfluencers steer clear of overt political advertising and tend to wrap political promotion within lifestyle content—for example having a meal or going on a morning jog with a local politician. At the same time, there are prominent digital influencers with established Twitter accounts and YouTube channels who are emphatic and direct about their political leanings.[100]

Through the Shailesh Ki Duniya (Shailesh's World) InstaReel and YouTube Shorts handle, Shailesh, a microinfluencer who traverses tier 2 and tier 3 cities like Raipur and Korba in the state of Chhattisgarh, advertises restaurants and colleges in an implicit manner while, for instance, taking a tour with his wife for an authentic Chhatisgarhi lunch or learning at a computer science lab. A similar strategy is adopted for the local member of the Legislative Assembly, Vikas Upadhyay, from Raipur West. Publicity videos become more about a day in the life of Vikas or Shailesh and his wife spending time with Vikas and his partner cooking in the kitchen or gardening bitter gourd (*karela*), Vikas and Shailesh hitting the gym, and Shailesh serving (*Sewa*) his constituents.[101]

There is often a tendency in distribution studies scholarship to suggest that the user-generated content moving through instant messaging platforms (WhatsApp) and short video platforms (TikTok, Moj) is less intentionally directed and hence is more specifically about circulation and less about distribution. When I have pointed out that political parties in India, using their cadre and local organizers, clearly direct their outreach messages through WhatsApp toward particular voters in specific constituencies (having gathered data using membership drives and the "missed call" initiative), I am told that is not really distribution—this despite political parties nowadays using influencer marketers and local microcelebrities (and even AI deepfake technologies) to personalize and customize their messages with specific sociocultural tells of a region where the mobile phone user, a potential voter, lives.[102] This has parallels with the language of customization and personalization that I hear from Netflix and other streaming company executives when they discuss knowing their audiences and subscribers, although with different intentions in mind. The entanglements between political, cultural, and industrial organizations have to be considered while studying and theorizing about distribution practices.

So far, we have discussed mobility of content; the next two chapters examine the circulation of digital money via mobile phones. The addition of neomobile users has not only lent new vitality to short video platforms in India but has also made it possible to include a wider population of the country in digital payments, lending, and banking initiatives. The implications of this expansion are discussed in the next two chapters.

4 Phishing: Insurgent Trajectories

This chapter investigates mobile phishing scams, with a particular focus on understanding how such scams reveal cracks in the new infrastructures of circulating digital money through e-wallets and mobile phones in India. I begin by examining the kinds of discursive arguments and infrastructural designs advanced to usher in digital money circulation in India. "Universal mobile access" is one of the nine pillars of the Digital India mission, which also had three core areas of focus: "digital infrastructures," "governance service," and "digital empowerment."[1] This suggests that mobile phones have become central to the biopolitical governance of the Indian state through online bank accounts and the unique identification system (UID or Aadhaar card). The free bank accounts that the government instituted ([PMJDY]—Prime Minister's People's Money Scheme) were to be authenticated through Aadhaar with the mobile phone. This Jan Dhan-Aadhaar-Mobile (JAM) payment pathway became the preferred infrastructure for welfare payments.[2]

The JAM vision was a way for the government to encourage its citizens to enroll in Aadhaar, open a bank account, and possess/use a mobile phone. Introduced in 2014, JAM was considered at once to be a neoliberal governance strategy as well as a way of espousing a particular model of digital financial inclusion. It sought to formalize the Indian economy, steps toward which were also taken through the widely discussed 2016 "demonetization" ("cashless agenda") move of the Narendra Modi government.[3]

Beyond the JAM trinity, the Government of India offered the Unified Payments Interface (UPI), and with demonetization, fintech start-ups and companies mushroomed, offering a variety of financial services from payments and loans to savings and insurance, all of which could be done through mobile phone apps, with a click, swipe, or tap.[4] This notion of "instantaneous customer satisfaction" where the state and its services could be summoned by the citizen at their convenience became a feature of the efficient administration and the "start-up state."[5] I begin this discussion with this elaboration of Digital India and its relationship to an intended formalization of the Indian economy and an intended financial inclusion of the "unbanked" poor in India to tell a different story of some of the unintended, or rather unpredictable/unforeseen consequences of the JAM and Digital India missions. This is the story of mobile phishing (or "vishing" [voice-based phishing]) scams involving credit cards frauds and mobile money circulation in the Jamtara district of Jharkhand.

In this chapter, I argue that JAM/Digital India missions created alternative possibilities for mobile money circulation, thereby aiding in phishing/vishing scams. Furthermore, I contend that credit card fraud and cybercrime are constitutive of the so-called financial inclusion schemes and other efforts by the government and corporations to digitize the Indian economy. The big political motivation behind the decision about demonetization and digital/mobile money was to formalize the Indian economy so as to stop the illegal circulation of "black money." One of the key campaign planks on which Prime Minister Narendra Modi's government had been elected in 2014 was to uproot corruption, which included a fight against black money.[6] Toward the end of the chapter, I discuss the ways in which black money circulated in the past and the processes through which "mobile money," or "digital money," circulates today.

The demonetization move, in particular, led to the rise of mobile wallet services such as Paytm and PhonePe as people were forced to adapt to digital payment methods to buy groceries and vegetables from street vendors. Because of its QR code–based payment acceptance solution that required neither a credit card nor mobile internet on smartphones, Paytm became popular among street vendors and small businesses who used unbranded feature phones.[7] Paytm had initially started as a firm providing prepaid mobile recharge services and added e-wallet services in 2014. By the time demonetization happened in 2016, Paytm was already way ahead of its competitors: This early mover advantage helped the company double its user base during demonetization. Such has been the rise of mobile wallets that some scholars have claimed that about 90 percent of digital financial transactions in India happen through nonbanking institutions like GooglePay, Paytm, and PhonePe.[8] The rise of the payment and money transfer infrastructure built by e-wallet services is significant because (as I discuss in this chapter) they are a key part of scam money distributions and reroutings in India.

Credit card frauds are being undertaken by teenagers in Jamtara, a rural district in the state of Jharkhand. We see one such scenario play out in the Netflix show *Jamtara: Sabka Number Aayega* (2020), in which a village youth uses a basic phone to make the call as his teammate works to open an e-wallet on a smartphone. Once the card details of the individual are entered, the transaction mostly requires a one-time password (OTP) to authenticate. The OTP is needed to transfer the money from the credit or debit card to the fake bank account (e-wallet) opened with fraudulent KYC (know your customer) documents. Here, the link to some of the authentication efforts through Aadhaar (that is, eKYC) cannot be missed. Authentication can now happen through the phone via instantaneous clicks, but that also creates room for fraudulent authentications and identifications. The duped urban customers fall for these scams because they are increasingly becoming used to having their troubles (with, for instance, an expiring ATM card) sorted instantaneously with a phone call, something that the state and corporations have sold as a convenience of living digitally in India.[9]

I analyze two seasons of *Jamtara: Sabka Number Aayega* (2020 and 2022) to understand how the show depicts the impersonation activities of scammers making phishing calls and represents the distribution infrastructure of money being siphoned into difficult-to-trace e-wallets and bank accounts.[10] While the Netflix show is fictional, the script was developed based on extensive research on the everyday lives and media practices of Jamtara's scammers, and part of my analysis in this chapter is based on interviews with the show's scriptwriters and research team. In the summer of 2021, because of COVID restrictions, I interviewed local journalists and denizens working and living in the districts of Jamtara and Ranchi, the latter being the capital of Jharkhand, over the phone. In July of 2024, I conducted an in-person fieldwork visit to three locations in the Jamtara district—the villages of Narayanpur and Karmatand and the district town of Jamtara.

With the proliferation of digitization and financialization amid neoliberal globalization, scams are constitutive of capitalism. In manipulating (or pirating) the operative logics of the telecommunication and digital finance infrastructures, Jamtara's scammers, like Jamaica's lottery scammers or Nigeria's "419" fraudsters, participate (or insert themselves) in "domains of capital and technology from which they were often thought to be excluded before."[11] Jovan Scott Lewis explains that mobile phone technologies (VoIP apps and VPNs) acted as new frontier spaces for Jamaicans undertaking the lottery scams, who lived amid the confluence of poverty and aspiration in Montego Bay.[12] Such digital spaces where they could appropriate capital were previously perceived to have been inaccessible to them. While the crew from Jamaica (Montego Bay) would impersonate US credit claims commission officers and make international calls to US citizens, the Jamtara scammers' calls were made to Indian cities, pointing to the rural-urban geographic and socioeconomic inequities in India. That such scam calls from Jamtara or Montego Bay even work (that is, are considered believable) reveals much about the mundane ways in which the procedures of call centers have been normalized and how contemporary digital capitalism creates monetary (in)securities among both the wealthy and the poor.

While the call center angle is a similarity between Jamtara and Jamaica's scammers, a striking difference in the case of the Jamtara story is that these scams targeted not transnational foreigners but fellow citizens in India. The call center industry in India's metropolitan cities mainly involved telemarketing and debt collection activities, but around 2011, some call centers started getting associated with scams and impersonations.[13] Such phishing scams involved call center operators in cities like Pune or Gurugram impersonating Social Security officers calling people in the United States as they worked from old data left behind in India by offshore companies. The incidences of cybercrime today, like the one in Jamtara, are no longer mostly about targeting foreign nationals but are directed toward urban Indian citizens, and the calls no longer originate from a call center in Gurugram or Mumbai but come from rural hinterlands.[14] Cyber criminals no longer operate out of cybercafes or call centers, but work from open fields with mobile phones. Being on the move on a motorbike or calling from the shade of bamboo huts or dry forests bordering the villages (of Karmatand and Narayanpur in the Jamtara district) helps the rural scam callers to evade being detected by cybercrime investigators who try to monitor cell tower signals and communications. So, rural landscapes play a role in the scam media infrastructure.

Studying mobile phishing scams in Jamtara therefore inevitably involves understanding how scams play out amid the rural-urban divides (and continuums) in terms of infrastructural connectivity and uneven regional and socioeconomic mobilities. While the real-life scammers live in the

Jamtara district, they do not operate from Jamtara town per se but from two villages, Karmatand and Narayanpur, about seventeen kilometers from Jamtara town. *Tand* means fallow land or infertile soil. The region lacks industrial growth, and without rainfall, agriculture suffers, so historically, there has been both unemployment and underemployment. While Jamtara certainly does not have internet bandwidths comparable to those in Delhi or Mumbai, its denizens today are connected to Delhiites and Mumbaikars through mobile phones, even as some of these connections are based or predicated on scam calls. What then are the consequences of such connectivity where erstwhile marginalized groups like the scammers of rural Jamtara are in possession of novel affordances for authoring and distributing information, whether it is phone numbers, SIM cards, or bank account details? In this chapter, I will provide a granular analysis of the local issues involved in Jamtara and its mediated connections to rural and urban India that entangle discourses of aspiration and cybercrime.

While the focus of much of this chapter is on mobile phishing scams in the region of Jamtara, and the Netflix show *Jamtara*, such scams are not exclusively based in Jamtara. There are lots of other Indian villages in many regions where such scams are being undertaken. Another much-discussed area of cyber-fraud operations in India has been the region comprised of Mewat, Bharatpur, and Mathura, which forms the tri-junction of the states of Haryana, Rajasthan, and Uttar Pradesh, respectively. This region has become a hot bed of scams because it allows easy passage from one state to another, and moving from one state to another makes it more difficult for the authorities to obtain the scammers' cellular data records and thereby harder to trace the culprits.[15] Jamtara has a similar geographical location. Located in the state of Jharkhand, Jamtara is proximate to the states of Bihar and West Bengal, and such proximity allows scammers in Jamtara to easily move into these states in order to elude authorities who are chasing them. These aspects of cellular signals and data, state borders and travel, add conjunctural moments of the convergence of transportation modes and communicative (mobile) media to the conceptualization of mobilities (un)-limited, a term that I have been theorizing about throughout the chapters of this book. Furthermore, I theorize about mobilities (un)limited in this chapter in relation to the aspirational politics of scamming as depicted in the Netflix show *Jamtara* and from my own on-the-ground observations in the Jamtara district.

I will begin with a discussion of the ways in which the phishing/vishing activities were carried out in Jamtara, and then explain the links of these activities to the Digital India mission, especially the distribution of mobile money. In the second season of the Netflix show, we see Jamtara-based scammers constantly needing to improvise on their digital conning methods, as the people they call seem to be in a better position to suspect their intentions than before. This leads to a multiplicity of scam cultures as novel scamming methods need to be devised in order to counter predictability. Amid a proliferation of scams in relation to cybercrime worldwide (and not just in Jamtara), such as lottery scams, advance fee scams, tech support scams, credit card scams, pig butchering scams, and internet romance scams, I examine how scammers and scammed victims perceive such scams. Should online/mobile phone–enabled scamming be considered crime or enterprising labor or just a side hustle? What kind of platform labor is involved in scamming?

Scammers themselves might be unwillingly recruited and forced by organized networked criminals to defraud people, as journalistic investigations by Al-Jazeera in the southern coastal city of Sihanoukville in Cambodia have revealed.[16] Do the unemployed youths in Jamtara turn to scams aspiring to a different lifestyle or simply in desperation after finding no other avenue of employment, and hence submitting to the allure of scam leaders? Especially in the second season of *Jamtara*, we see all kinds of youths, even school children, being recruited to participate in scamming activities. Toward the end of this chapter, I discuss some of the implications of such scamming labor regimes and their relation to the distribution geographies of mobile money.

Scam Calls: Impersonation, Convenience, and the History of Call Centers

Jamtara is the name of the district in the state of Jharkhand as well as the town that serves as the district's administrative headquarters. Youths from this district impersonate bank officials and call urban customers to inquire about their card details. They explain that they require those details to offer the valuable customers specific rewards or stop their ATM card from being blocked. Such credit card scams based on vishing have increased in frequency. Sometimes, customers hesitate to share card details and are not fully convinced but then finally relent, lured by the convenience of being able to continue using their cards or by the promise of gifts from the bank for being a "valuable customer."

A scenario plays out early in the second season of the Netflix show *Jamtara* when somebody receives a scam call mentioning that their paycheck could not be deposited because of some verification issues with their bank account. The person has the option of either divulging identity/account information or taking an arduous journey to the bank. Given this situation, there is a strong likelihood that the customer/consumer might end up selecting the convenient option of spilling identity information and account details on the phone. Phishing operations depend on not just offering convenient options but on winning the trust of to-be-duped customers on the other end of the mobile phone call. This ability to have customers place their trust in scammers and divulge personal information requires, to begin with, some vital information (data) about their whereabouts or vulnerabilities. For example, in the above-mentioned case, the information that a specific person has wages deposited into an account in a particular bank is crucial knowledge. Certainly, a convincing imitation of a type of person or institution brand (bank official/bank) they can trust is also very important. Further, at various points in the show, it is indicated that Jamtara scammers are using call center data records to call potentially vulnerable victims/customers who might take the bait more easily or who have significant bank balances. Rinku, a new character introduced in the second season of the show has more success in scamming because he knows a mole in Noida working inside the call center, who discloses data about premium account holders, bypassing security protocols.

To get the attention of their targeted victims (or customers) and enhance the believability of their con script, Jamtara's scammers impersonate the procedural formality of call center agents. This is something Jovan Scott Lewis also mentions about Jamaica's lottery scam group, who also evoke the professionalism of customer service providers in their interactions with their supposed lottery recipients. Throughout the late 1990s and early 2000s, call center labor was heavily outsourced and offshored from United States to both Jamaica and India during the opening up these two countries' economies.[17] It is not just the history of call center development that acts as a backdrop for the material media practices of these scammers. Both the Netflix show creators and Jamtara town's locals told me that they believe

Figure 4.1
Calling from bamboo huts: Phishing operations as depicted in Netflix show *Jamtara*. Courtesy: Digital18.

somebody from Jamtara who worked in a call center in Delhi or Mumbai came back home to Jamtara and then started this scamming project. Lewis's Montego Bay–based scammers also noted that they "learned the 'the game' from co-workers" at a call center.[18]

During late 1990s to early 2000s, global multinational corporations chose India as a destination for outsourcing work. The country witnessed a huge increase in call center jobs, with easily available cheap labor proficient in speaking English. That said, over time, the Philippines and several Latin American countries, including Mexico, overtook India in call center jobs. This is because of a number of reasons: time zone considerations, other countries being able to attract outsourcing opportunities and investments, "neutral accents," and a nationalist racist backlash against Indian call center operators in the United States.[19] By 2011, a segment of the call center industry turned to scams, in part because the legitimate call center industry did not seem to grow any further, while cheap labor was available and waiting (in the form of unemployed English-educated youths). Even more importantly, Winifred Poster writes, "the Indian call center industry has been the recipient of a decade's-worth of training on how to carry out overseas digital communication."[20] Such training involves not just conversation skills but also training in "producing fakeness" so as to mollify distraught American customers, which included sending consultants from the United States to train employees in India regarding the minutia of "outsourced voice work."[21] Scamming call center companies operated in plain sight, sometimes amid (and adjacent to) legitimate call center firms, sharing the same glitzy high-rise buildings in information technology parks such as Cyber City near Delhi or Sector V in Salt Lake City of Kolkata.[22] Migrant youths from small towns in India went to cities and worked in such call centers, sometimes not even initially knowing that the call center employing them was engaging in deceptive scamming practices. Here, the training from US consultants in impersonating specific accented voices and persuasive intonations was passed down to new employees, and somebody, as the legend in Jamtara goes, who worked in one such city-based call center moved back to Jamtara to then share these skills with his colleagues (or gang members).

Jamtara's scammers, even men, are depicted as often taking on feminine voices and tones to persuade their customers. The gendered aspect of voices and the affect they convey are central to call centers, and Winifred Poster has remarked on the "rise of communicative capitalism" with the increasing importance of "sound" and "worker vocalization" in maintaining customer relations and "measuring productivity of service economy."[23] As Jamtara's scammers are shown to mimic call center vocalizations, tones, pitches, and rhythms in order to dupe urban elite customers, there is a satirical praxis at play, which the Netflix show's audience can certainly recognize. The call center and business process outsourcing phenomenon therefore plays a role as both a historical backdrop to contemporary phishing/vishing activities as well as a material means of skill transmission of impersonation and misrepresentation practices to the Jamtara scammers calling from rural locations. These impersonation/misrepresentation tactics create a recognizable and familiar space of proximate interaction in the minds of their urban elite customers, who are used to hearing call center voices. However, this impersonated proximity is subversive, and even possibly insurgent, and tries to dismantle the unevenness of rural-urban divides.

The *Jamtara* production team told me that they thought about developing the "research concept" for the show when they read Deepu Sebastian Edmond's December 2015 report about phishing in the district of Jamtara in *The Indian Express*. Edmond was taken around Jamtara by subinspector Nitish Kumar, who stopped at the Airtel tower in Kala Jharia village and noted, "We currently have the capability of monitoring only one cell tower at a time. The Airtel cell tower outside Kala Jharia village sees 3,000 outgoing calls a day, almost all of them to other states. It should be 700–800 at a normal tower."[24] In the vacuum of joblessness that exists across vast areas of India, and that has existed in Jamtara district for a long time, the 3G cell tower made its entrance and fueled aspirations and mobile phishing (figure 4.2). With Jamtara district becoming the origination point for a lot of vishing calls, the regulatory state over time decided to have a police station dedicated to cybercrime in the heart of Jamtara town alongside the old courthouse (figure 4.3).

With the expansion of digitization into India's economic periphery, both the aspirations of locals and the rates of cybercrime have increased. Growing up in the Jamtara district, where there is massive unemployment (and underemployment), the mobile phone scammers have had incomplete childhoods. From what I could gather from my interviews with locals (some conducted on phone in July of 2021 and others undertaken during an on-site tour in July 2024), these teenaged scammers, often referred to as "cybers" by the locals, grew up too quickly, as their education was shortened and they were compelled to earn money. Some of these youths are savvy at impersonating voices on the phone and in gathering multiple SIM cards and mobile phone numbers for the group's phishing (or vishing) operations. They have used the money gathered from such phishing operations to construct houses and for other businesses, including setting up coaching classes.

In the Netflix show *Jamatra*, these young boys are sometimes introduced as *daswi pass* (tenth standard pass) or *panchwi fail* (fifth class fail). The show depicts the homosociality of this group of boys as they idle around tea stalls or play video games on phones or plan the next mobile phishing attack: gathering a list of phone numbers to call, strategizing about how to con people into believing them, and devising ways to evade the radar of cybercrime investigators. These *timepass* sessions involving mobile phone calling might be construed as cybercrime by some but as entrepreneurial cultures by others. This ambivalence exists in a context where the Indian ruling dispensation is touting its Digital India mission and promoting the growth of mobile phones, e-wallets, and online bank accounts even as unemployment increases with a labor oversupply and unmet expectations, especially among the youth demographic. The youth with a mobile phone is an aspirational subject in the state's calculus, but the state recognizes that

Figure 4.2
Cell towers in Karmatand, Jamtara. Photo taken by author.

Figure 4.3
Cybercrime Police Station (Jamtara). Photo taken by author.

the youth is also a vulnerable category at the threshold of adulthood, prone to social disruption.[25]

The mobile phishing scammers in Jamtara district are called "cybers." Partly, this name has to do with their association with cybercrime. Another reason is that an earlier iteration of the vishing activities was conducted from cybercafes. With internet available on 3G and 4G phones, nobody needed computers to conduct the monetary transactions involved in phishing calls. A few years ago, wrote journalist Pankaj Mishra in 2018, mobile phishing gangs used to operate from one of the dozens of cybercafes in and around Jamtara, but over time, the cops have made that difficult, with cybercafe owners being co-opted into police informer networks. Now "the phone" is "their cybercafé."[26] Mobile internet makes it easier to transfer money to e-wallets, not just Paytm and Oxigen, but also Tapzo (one of India's first super-apps—integrating multiple services together before it was bought by Amazon) and TMW. Here, we see the scam itself transform as ICT4D (information and communication for development) discourses morph into M4D (mobile for development) visions. The mobile phone is central to the Modi government's Digital India vision.

The mixed success of rural PC kiosks and telecenters, including ITC's e-choupals, has been well documented by several scholars of Microsoft Research India (including Nimmi Rangaswamy and Kentaro Toyama), and they have identified a variety of socioeconomic barriers (including class and caste) as well as confused branding (from farmers seeking agricultural advice to children playing computer games) and difficulty in meeting frequent maintenance needs.[27] That said, the computer equipment at the government service centers, or *kendras*, at each *panchayat* (village council) over time have proven crucial to connecting digital bank accounts, the unique identification (UID) system, and the mobile phone. This bundling of these three elements has been vital for the Digital India mission, based on the belief that every person in India will at some point have a mobile phone, if not multiple mobile phones.

Alternative Circulatory Pathways for Digital Mobile Money

Part of the Jamtara story is about impersonation—calling and scamming as performance. Another part of this scamming story is about mobile money distribution. And this has profound implications for rural media infrastructures and what is happening in India with the Digital India mission. These youths would rent out the new Jan Dhan accounts from the elderly in Jamtara (and adjoining villages) for a sum of five thousand rupees. These Jan Dhan accounts would be where the youths would transfer money from the duped urban card holders' accounts. In the Netflix show *Jamtara*, the lead youth scammer, Sunny, is seen accompanying an elderly uncle to the bank to withdraw cash, and the bank teller remarks that he saw Sunny come in with another uncle yesterday (figure 4.4). Sometimes the account holders' thumb impressions are taken on plain paper and are then scanned digitally.

One of the ways in which police officials from other states broke up the phishing racket was when they found money from the credit accounts of Indian citizens living outside Jharkhand going disproportionately into Jamtara residents' Jan Dhan accounts. From 2016 to 2019, well before the Netflix series, many news reports discussed how money from a credit card fraud or the Unified Payments Interface was illegally transferred from cities like Delhi and Bhopal to Jamtara's PMJDY accounts. When police teams

Figure 4.4
Sunny accompanying different elderly people to the bank to collect money transferred to their accounts. Courtesy: Digital18.

from other states arrived in Jamtara to trace the account holder, they would be surprised to discover that the accounts, mostly owned by the elderly under the PMJDY scheme, had been shared with scammers in return for Rs 5,000 per month. Sometimes the money from those accounts had already been withdrawn using duplicate thumb impressions.[28]

Jhan Dhan Yojana literally means People's Money Scheme. The Jan Dhan accounts were an initiative of the Indian government so that the previously unbanked would have bank accounts. Credit card fraud and cybercrime are constitutive of the so-called financial inclusion schemes and other efforts to digitize the economy. While the scammers used the Jan Dhan accounts to disturb the logics of the digital financial economy, it also needs to be acknowledged that the vishing/phishing youths were finally apprehended by the cyber police because they could see the digital financial records of money transfers to bank accounts in Jamtara. This digital trace thus is a key feature of cybercrime forensics. Here, I should add that Jan Dhan accounts were certainly not the only accounts to which money was being transferred:

As I noted earlier in the chapter, after the initial demonetization drive, several fintech companies emerged that made mobile payments and transactions so easy that Jamtara's scammers had many options for a variety of e-wallets, including Paytm and Tapzo. Also, it needs to be noted that the mobile money was often successfully rerouted through several e-wallets, making it difficult for the police to trace. Similarly, with multiple SIM cards being used and then discarded, it was difficult to track calls back to the impersonator or connect a particular SIM card to a specific e-wallet. Many local Jamtara journalists maintained that most of the cases that the Enforcement Directorate successfully filed against scammers in Jamtara were based on them having acquired disproportionate assets (multiple motorcycles, refrigerators, immovable properties, and TV sets). Scammers were simply unable to explain how they had the money to buy such high-value consumer goods.[29]

A closer look at the implementation of Jan Dhan Yojana, or this coercive encouragement (or nudge) by the government to get people to have Jan Dhan accounts, reveals its own "mediatic" apparatus entangled with the implementation of the Aadhaar (UID) project. The *kendra*, or government service centers, one of which exists in every panchayat in Jharkhand, is a crucial mediator in the Digital India infrastructural assemblage. At one level, in ICT4D parlance, it can be described as a PC kiosk or even a cybercafe, but most importantly, it functions as a delivery center for services, where villagers can apply for caste, income, birth, and death certificates. Often operated by intermediaries between government bureaucrats and village locals, private sector services such as mobile recharges are also available in some of the *kendras*.[30] These *kendras* have played a very important role in the Digital India mission, whether for the Aadhaar card or Jan Dhan Yojna, and the two are related quite explicitly. One of the policy measures implemented is that instead of welfare benefits in kind like food grains or kerosene oil, social stipends take the form of direct benefit (cash) transfers (in cashless form initially) to these Jan Dhan bank accounts. This is to enable "the poor" to access benefits at their doorstep, or in their village, by using their Aadhaar, biometrics-based identity numbers.

There has been a concerted push, as India's leading labor journalist, Anumeha Yadav, writes, to drive the "JAM" mission, which is the bundling together of Jan Dhan bank accounts, Aadhaar identity numbers, and mobile networks by the government as the cornerstone of its welfare policy. Both strategic and tactical coercion are at play here: Particular pension schemes for the elderly or scholarships for children that were available through other means have now been channeled through the Jan Dhan account so that if you do not open a bank account through the Jan Dhan Yojana, your son or daughter cannot receive the scholarship (or you may not receive your pension).[31]

Priya Chacko notes that the Pradhan Mantri Ujjwala Yojna (Prime Minister's Brightness Scheme [PMUY]), which provides cooking gas connections to poor families, requires recipients of the scheme to have bank accounts through PMJDY.[32] Part of the reason for such interlinking between PMUY and PMJDY is because the PMUY offers a one-time subsidy for a liquid petroleum gas connection, with recipients encouraged to take out a loan to buy a gas stove. The oil marketing companies providing such loans recover them in the form of gas refills paid by the recipients. PMJDY is thus crucial if Indian citizens want to benefit from the subsidies, loan programs, and cash transfer schemes offered by the government. The PMUY scheme is promoted with the slogan "mahilao ko mila samman" ("respecting the dignity of women") and, along with PMJDY, can be seen as a way through which government-mediated mobile money distribution becomes part of empowerment and financialization projects. These projects serve both political and neoliberal logics, transforming the Indian population into what Priya Chacko calls "virtuous market citizens."[33] Political scientist Christophe Jaffrelot has argued that the Jan Dhan Yojana was neither a redistribution scheme nor did it solve the problem of inadequate access to formal credit, as the urban and rural poor continued to depend on moneylenders.[34] A significant number of accounts, after being opened, remained inactive, but a large segment of population did monitor their accounts, because that is where the direct benefit transfers started appearing. Jaffrelot contends that the Jan Dhan Yojana was never meant to provide India's poor with money, but it endowed them "with a sense of dignity and recognition" similar to some of the BJP government's other programs, like Swachh Bharat Abhiyan (Clean India Mission).[35]

I interviewed Jamtara locals, journalists covering Jamtara, and a TV crew that covered the subject of vishing in Jamtara. Some of them viewed Jamtara's boys who indulged in mobile phishing as entrepreneurs and part of India's aspirational youth, while others suggested that the story of Jamtara's boys is the story of changing crime: While earlier these unemployed/underemployed youth would rob passengers in train, now they were scamming customers on mobile phones.

A Karmatand village resident I interviewed by mobile phone (in July 2021 when I was in Ranchi, but because of the recent second COVID wave, the interviews were conducted on phone rather than in-person in Karmatand) told me, "Gaun mein sab waise mobile use nahi karta hain" (Not everybody uses mobile phone like that in the village).[36] This statement suggested two things: (1) The villager did not want to be associated with the scammers, and (2) he was acknowledging that he was not savvy at using the smartphone for calls and monetary transactions.

During my July 2024 fieldwork visit to Jamtara, I found myself chatting with a teenager who was helping his father run a *dhaba* (roadside café/food-stall). This *dhaba* was just outside the village of Narayanpur, a village that had earned a lot of ill repute for phishing scams. As I was waiting for *parathas* (flatbread), he asked me if I was on my way to Deogarh. This was *Srawan mas* (month of Srawan), and usually during this time every year devotees go on a pilgrimage to pray to Lord Shiva in a temple in Deogarh. The *dhaba* was on the way, and people would stop to eat lunch and rest for a while before proceeding to Deogarh (see figure 4.5).

Figure 4.5
Roadside motel on the way (close) to Narayanpur, Jamtara. Photo taken by author.

We started conversing, and after some time, the topic moved to phishing scams and Jamtara's reputation. This teenager told me that he knew some people in his village who indulged in such *dhanda* (business), but he had never dabbled in such things.[37] He kept stressing that he is learning various skills—how to cook tea and how to check the air pressure of truck tires—to be able to expand his father's business at the food stall. He gestured to some construction work that had started on the second floor, explaining that they are hoping to put together a motel so that travelers who stop for food and drinks might consider staying there if they need to. He hastened to add that the money for the second-floor construction of the motel was not coming from phishing activities but from selling some agricultural land that his family owned: "Amra jomi bikri kore dooh tola banachi" (We sold land to build the second story/floor).[38]

These tales of dissociation from the business of scamming and of wanting to follow other entrepreneurial routes like restaurant and hotel businesses that I heard in Jamtara suggest that locals in this district do not want to all be painted by the same brush. They want to fight against the perception that Jamtara is the scam capital of India even as they acknowledge that despite the risks associated, for some, leaving the *dhanda* (scam business) is not that easy: "Jiska dhandha ek baar ho gaya, woh nahi chodega . . . hum nahi hain usmein" (For those, whose business it has become, they will not leave that business (scamming) . . . I am not in that).[39] This is primarily because if the scam works out, there are high rewards. There is also a recalcitrance to such scams, as scammers do not stop after being caught or if their one scamming method has been discovered by the cyber-police. They devise new scams ranging from fake lottery awards to misleading work-from-home jobs. This continuation of phishing as *dhanda* can be seen as subaltern recalcitrance that continues to defy state authority by exploiting the porosity of erstwhile sealed/closed circuits, whether it be the call center system or the digital banking system, that seem now to be prone to leakages.[40]

A local journalist explained the internal heterogeneities regarding mobile phone usage, noting that every panchayat in the district of Jamtara or in rural Jharkhand might have about fifteen smartphone users, and the kinds of phone use as well as the information or misinformation climate in a place was dependent on how these fifteen smartphone users influenced the behavior and knowledge of the rest of the villagers. For this leadership of smartphone holders he used the term "community mobilization" in the following statement: "Along with mobility, we need community mobilization . . . every panchayat has 15 smartphones but that is it."[41] One of the reasons for the delay in catching Jamtara's mobile phishing scammers was the difficulty of tracing call records, especially because Jamtara is a border district, and the scammers would move between Jharkhand, West Bengal, and Bihar to evade police interception. This is something future cybercrime investigations need to wrestle with, as noted by a cybercrime police person in a research interview: "Within any state, within a district, within a police station, there is a body within a jurisdiction . . . this is what happens when you move from different state . . . cybercrime police will have difficulty accessing CDRs (call data records)."[42]

In corporate and political brand promotions, the mobile phone as a technological artifact has become increasingly associated with aspirational cultures of Digital India: A symbolic association with the phone in a citizen's everyday life is necessary in presentations of the "technological self" working toward development and entrepreneurship in contemporary India.[43] So, for some Jamtara boys, vishing activities are part of *dhanda* (business). During my fieldwork in the Jamtara district in July 2024, I found that in a freshly painted wall ad, Airtel was announcing a new cell tower in Karmatand, perhaps tacitly acknowledging that telecom service was a key part of the economy of the village, a village widely associated with mobile vishing scams (figure 4.6).

My point is that an attention to media rurality in relation to the Digital India mission (involving *panchayat kendras* [*pragya kendras* in the state of Jharkhand] and the Jan Dhan accounts) will suggest that it is not just in such brand promotions but also in concrete media practices that the mobile phone is being made part of a prescribed ideal entrepreneurial citizen's life.

It is difficult to side-step discourses of "lack" or "divide" (whether it is the digital divide or urban-rural divide) while discussing media rurality in India, so entrenched is a particular kind of ICT4D discourse. The mobile phishing scams do not so much expose the contradictions and shortcomings of the grand connectivity projects like Digital India. Rather, the phishing activities suggest that the way mobile and digital connectivity will play out in terms of rural-urban relations remains contingent and unpredictable. The scam calls exploited the idealizations of instantaneity, immediacy, and convenience that have become the promises, values, and enactments that the Indian state and corporations are promising. The different uses of Jan

Figure 4.6
The Airtel wall ad reads: "Airtel's new tower, now in Karmatand." Photo taken by author.

Dhan accounts and the ways in which scammers created alternative monetary distribution systems indicate that informal piratical practices cannot be eschewed away from the formalization of finance through digitization. The vishing scam also points to how infrastructural mediations in rural areas can create new entanglements with and alternative circulatory pathways for urban media formations.

Aspirational Trajectories: Caste and Gender

While many attempts at duping people might fail, it is always possible to procure a large catch of money from one very successful attempt at digital/mobile scamming. This is precisely why scamming can be a "lucrative career choice . . . very profitable career choice" as noted by journalist Snigdha Poonam, who has extensively interviewed scammers in India.[44] Instead of earning a measly salary every month from a substandard job and feeling stuck as the Indian dream unfolds frustratingly slowly, "one successful scam could land 2-3 lakhs."[45] The Jamtara story as shown on Netflix was a way to depict a new kind of crime (that is, cybercrime) and interrogate the motivations for engaging in such a crime. What kind of aspirational mobilities were engendered by the possibilities opened by mobile phishing operations? Could the influx of money help "lower-caste" scammers to finally rise up and confront the "upper-caste" local politicians as shown in the Netflix show?

During his research for *Jamtara*, one of the concept writers, whom I will refer to as N., found the politician figure (the local member of the Legislative Assembly [MLA]) to be a tertiary character. In his initial script, he suggested that the MLA's role be limited. N. had felt that the kind of protection the MLA was offering to the Jamtara scammers was very "soft" protection, which would involve making one arrest and letting eight others go scot-free, making sure that the district's name was not sullied and mobile phishing could continue, as it was the primary economy of the region. N. was keen to figure out what kind of crime cybercrime is and found that the overwhelming presence of the politician protagonist that the series producers later inserted compromised the procedural intricacies of how mobile phishing actually operates. N. was interested in emphasizing details of how scamsters made calls from bamboo shrubs and on-the-move motorbikes to evade the scrutiny of investigating authorities:

> The story is about the crime we are trying to figure . . . yeh crime hain kya. Yeh pakad mein kyun nahi aa raha hain [what is this crime? Why is not being caught?] . . . the stupidest thing that they did was that Brajesh Bhan Bhayya created a call center in this house . . . stupidest thing . . . because if you do that and you are calling from there, then your cell tower information will say that 50,000 calls are happening from there, then Brajesh Bhan Bhayya can easily get convicted, then the entire operation will stop. Why the operation works is . . . primarily because these guys are so particular about evading cell tower information, that is why they make calls from inside bamboo shrubs . . . they make calls while they are on bike . . . so between towers they are moving . . . now these are the information that are exciting.[46]

N. was happy to have been funded to do the research for the script but then found his creativity scuttled by the rest of the production team's decision to overdramatize the fight between the MLA figure and Sunny, which went against the research he had done for the script. N. further explained that "producers ka hain ki jo chal gaya, wahi banana hain [Producers just want to follow the trend] . . . Mirzapur chal gaya toh [Mirzapur worked and so] everybody wanted to do a Mirzapur copy. I wanted Jamtara not to be another Mirzapur but it failed."[47] N. believes that Jamtara went the route of *Mirzapur* (2018), a crime thriller streaming on Amazon Prime, set in the small town of Mirzapur in eastern Uttar Pradesh, which portrays a more conventional crime of mafia-led gangsters and gun violence, even though Jamtara's mandate was to represent cybercrime, a different kind of crime that did not necessarily involve guns.[48]

That said, director Soumendra Padhi seems to have been interested in how the influx of money could potentially disrupt the caste hierarchy between the upper-caste Bhan and lower-caste Sunny. At several points in the show, Sunny aspires to break out of the caste disparities with the aid of the colossal amounts of money he collects from phishing scams. This is an aspirational narrative that the Netflix show engenders, catering to both its local audiences (in small towns, who identify with such aspirations) and transnational audiences (who are interested in progressive values).

In various discussions related to *Jamtara*, I found that aspiration was both mentioned as part of the production narrative but also dismissed as being too universal or ubiquitous a theme to be preoccupied with. For some in the production team, the aspirational narrative in *Jamtara* was essential to appealing to wider audience segments. However, many journalists who had covered the region of Jamtara and several others in the *Jamtara* production team wanted to see the story represent not an aspirational narrative of youth but a moment of changing crime: those who used to rob passengers in trains were now indulging in phishing/scamming. In fact, the metacommentary about changing crime unfolds as a cold open introduction to the second season of *Jamtara*. One of two travelers journeying to Jamtara remarks to his friend that there used to be stories about ferocious dacoits in the dreaded forests and ravines of Chambal, such as Dadua and Sultana, who would kill anybody at the drop of a hat. His friend responds that those are bygone days: The new dacoits (*dakaits*) are from Jamtara and are a hub of "smooth-talkers" (*bakaits*).[49]

Continuing with my interviews with local journalists and the Jamtara production crew, some others noted that aspiration and changing crime were both fine but narrow optics. They had to be looked at within the context of the regional disparities, caste privilege and discrimination, poverty, and lack of infrastructure that marked Jamtara.

In the over-the-top (OTT) thriller *Jamtara*, the audience witness various aspirational stories. Sunny and his gang want to continue earning money through mobile phishing and investing it in constructing even bigger houses and maybe also in opening coaching institutes. Rocky is an aspiring political leader. Gudiya, who runs an English coaching institute partly financed by Sunny is shown to be saving money so that she can buy her way out of Jamtara and go to Canada. She refers to Jamtara as a "shithole" and is seen negotiating with travel agencies about how much a Canadian travel visa will cost her (figure 4.7).

Gudiya is not just ready to be business partners with Sunny in running the coaching institute (which provides Sunny with more workers to conduct fraud calls) but also marries him for money that she saves, making it clear to Sunny that the marriage is just a contractual/monetary arrangement

Figure 4.7
Gudiya's desire for mobility out of Jamtara to Canada. Courtesy: Digital18.

for her.[50] Purnima Mankekar has noted that "aspirations mediate how subjects form their worlds and inhabit and navigate them."[51] The disjuncture between Gudiya's present in Jamtara and her imagined future in Canada guides her actions. The last scene of the last episode has Gudiya running with Sunny with a bag of cash she has saved as Brajesh Bhan's goons fire bullets at them. Gudiya and Sunny are climbing a gate. It is the gate that separates them from Bhan and Jamtara, and the freedom beyond Jamtara, the beyond that Gudiya has always wanted and that Sunny has never understood why Gudiya wanted. Gudiya has thrown the bag containing cash to the other side of the gate and is about to climb when Sunny gets hit by a bullet.

I mentioned to journalist Deepu Sebastian Edmond that this scene suggests there is an aspirational goal that Gudiya is running toward and that an audience in the Indian diaspora (or cities) would identify with this aspirational migration. Edmond critiqued my interpretation and questioned the whole aspirational India idea. Edmond noted that people in the diaspora or the ones in cities like Bangalore observe Sunny and Gudiya from a vantage point of the raised observer and that this is precisely the problem because "aspiration" involves a "looking down" upon: "You have already made it. Everyone else is aspiring. My issue is with where the observer is . . . observer is at a vantage point . . . it is a crude word . . . looking down at those who are coming up, those who are wanting to come up."[52]

Journalist Pankaj Mishra, who covered the Jamtara story in a conversation with me, criticized the aspirational narrative as being created by foreign reporters from the *New York Times*: "Aspiration is . . . a way in which foreign media like NYT looks at India . . . all these terms become manufactured like that but at the core it is crime being redefined . . . you have to understand where they [mobile phishing youths] come, what is [sic] their families, there is no other occupation they do . . . there are deeper societal reasons."[53] Mishra notes how he had extensive conversations with families who had lived in Jamtara for a long time, and they explained how since preindependence, Jamtara was part of the "Jehar Khuran" belt where robbers would mix drugs in food and pass it around in railway coaches, eventually robbing passengers. The aspirational framing, according to Mishra and Edmond, glossed over the societal complexities of Jamtara. For on-the-ground journalists, the framing involving tracking shifting registers of crime was more cognizant of the societal problems in Jamtara, of their difficult economic conditions where the infrastructure of crime had shifted from railway tracks to cell towers. Having reflected on these interviews, I do not see why the "shifting crime" frame and the "aspiration" frame have to be mutually exclusive in any analysis.

The aspirational story of Gudiya is a relatable element in the transnational travel of the show through Netflix. Indeed, everyone aspires, but their aspirations are different and possibly gendered. Sunny wants to stay in Jamtara, aspires to have a big house and eliminate caste hierarchies through money. Gudiya believes only escape from Jamtara has transformative potential for her. She is indignant about the structural forces of patriarchy that often stop and accost her as she walks on the roads in the village. Spatial and gendered binaries remain in Jamtara, and Gudiya does not see herself affirmatively cohabiting those spaces: A move to Canada will be transformational. But there are risks in being poor, indignant, and educated and having such high aspirations. Her aspirations are transgressive for the anxious masculinity that Brajesh Bhan represents, and she finds herself at the receiving end of Bhan's sexual and physical advances.

Caste and gender thus not only influence but also recalibrate the potentials and limits of socioeconomic mobilities made possible by mobile phone–driven money scams. The mobile phone allows for a new kind of subversive proximity (even intimacy) of rural scammers and duped urban elites, which contributes to theorizations of mobilities (un)limited. The mobile money movements and reroutings into rural scamsters' e-wallets and Jan Dhan accounts across rural-urban divides and social spheres generates novel relations and transactional orders. However, if we go by the narrative of the Netflix show, there are some limits to the potential of these novel distribution geographies of mobile money to shape socioeconomic mobilities in the villages of Jamtara. At the point where the second season of *Jamtara* begins, the audience finds out that Gudiya, instead of leaping across the gate and escaping to Canada, tried to help Sunny get up after he was shot in his leg, and both of them were caught by Bhan's guards. Sunny is in a hospital slowly recovering from the injury, and Gudiya is imprisoned.

When she is released from jail, instead of pursuing transnational migration, Gudiya decides to stay in Jamtara and fight an election battle against Brajesh Bhan, and this way the political angle, already present in the first season of the show attains center stage in the second season. In the timeline of the show *Jamtara,* election preparations are underway just about when the demonetization move by the central government of India is explicitly declared. Elections in India, as is well known, are very much about money power, but with demonetization, there is more regulation of conventional money laundering and "black money" (*kala dhan*) operations, which creates the need for new ways of rerouting mobile money and dodging the authorities.[54] This works well for the show's narrative Gudiya and Brajesh Bhan compete to raise money, and a lot of this money arrives from multiple online scams, all enacted through mobile phones.

Multiple Scams and Labor Regimes

When Netflix dropped the first season of *Jamtara: Sabka Number Aayega* in early 2020, the streaming audience considered it refreshing because of a novel premise that suggested that the cybercrime capital of India was not a metropolitan city but an impoverished district in the rural hinterlands of a very backward state, Jharkhand. The mobile phishing operations were a new kind of crime not yet taken up in the genre of crime thrillers. When the second season arrived, the novelty factor had gone. Many newspapers and news channels had reported about Jamtara and reinforced its status in the minds of many Indian citizens as the online scam capital of India. To its credit, the second season of *Jamtara* did try to show that the scamsters were constantly improvising to develop new strategies, tactics, methods, and scenarios of scamming. Some of this depiction is realistic, because with the mushrooming of scam operations and India's digital citizens' increased awareness that they could potentially be tricked, scamsters could not successfully deceive unless they devised new ways to scam people.

Multiple online scams have emerged in India, as can be gleaned from reading news reports, and the second season of *Jamtara* presents several such scams, such as fake lottery awards, emotional blackmail about exam fees, misleading photos sent over dating apps for money, and misinforming about work-from-home jobs. Several of these have an element of the well-known "advance fee" kind of scams, where in order to receive a large financial reward (whether a high-paying job or a potential big loan or winning a big lottery), the (conned) customer must pay "perpetrator taxes or fees."[55] In the second season of *Jamtara*, Sunny recruits a group of students to be part of a massive lottery scam where people are called to be misinformed that they have won a big award from *Khel Karodon Ka* (The Crore Game), but in order to collect the money, they have to complete a formality that involves paying twenty-five thousand rupees in taxes as part of governmental regulation. Here, the believability factor is enhanced by the association with the *Khel Karodon Ka* brand, a famous game quiz show, with a nod to the actual TV show *Kaun Banega Crorepati* (KBC).

In another call, the Jamtara scammer poses as a work-from-home recruiting company caller. He announces that he has a data entry job that entails form-filling for the person on the other end of the line. However, to access the company portal and start the job, the job seeker has to deposit Rs 15,000 on a United Payments Interface (UPI) link sent to her. Here too, the believability factor is enhanced by the caller speaking in Marathi, which is the mother tongue of the person who is looking for the job. The phishing operator reassures her that if and when she decides to quit the job, the Rs 15,000 will be returned to her. Other scam scenarios, which are often discussed on Indian television news shows but are not shown in *Jamtara*, include sextortion involving leaking morphed and explicit images and SMS ultimatums about power disconnection at homes unless money is sent in the form of digital payments.[56]

Starting in May 2023, the work-at-home job scams in India often followed a set pattern where a private message was sent to the WhatsApp number of a person who was at that time looking for a job, as indicated in their LinkedIn profile. Somebody posing as an HR employee of a firm would contact the job seeker, and a web search would lead to a fake website, indicating that the company did indeed exist. The job offered involved tasks such as rating YouTube videos or commenting on films. The targeted person would be asked to join a Telegram group where they would be assigned tasks and initially rewarded for completing them with small amounts such as Rs 150. However, after this, the targets would be assigned prepaid tasks where they were encouraged to first pay amounts of Rs 500 and then Rs 1,000 and then Rs 5,000, all the time being promised that their initial investment would double or triple in returns. These directions provided as Telegram messages were timed to disappear. When the target complained that no money was being returned, they would be warned that if no more money was provided, they would lose previously deposited money.[57] Here, we see that chat apps such as WhatsApp and Telegram, which are often associated with spreading misinformation ("fake news") related to political matters and social conflicts (as discussed in chapter 2), were now being used to misinform people about money, financial matters, and jobs. The affordances of smartphones and digital platforms create new communication channels for financial exchanges and for distribution of financial products. Hence, there is the need to study the circulation of content and money—that is, news and finance—together: A recent conference poster suggested examining "misinformation in/and fintech (financial technology)."[58]

Toward the end of the second season of *Jamtara*, Gudiya and Sunny, in the face of Brajesh Bhan and his hired boys raising lots of money from scams, find themselves in a desperate situation, needing to concoct a money-making scam to remain competitive. Gudiya recruits volunteers from her party and Sunny enlists kids from middle school to participate in a labor-intensive scheme of deciphering the four-digit pins of stolen ATM cards from Brajesh Bhan's hired gang members.

Sunny has found particular apps that do not stop after one has input the wrong pin, so they can try multiple pin combinations for a particular card until they arrive at the right number. This may not be the most innovative plan, but instead of running automated hacking software to decode the pin, he has the gathered people do it collaboratively. So, each set of students and party volunteers is assigned to try out a certain number of pin combinations for a particular card: someone tries from 0001 to 0200, the next person attempts from 0201 to 0400, and so on. Once the right pin is arrived at, a new card is assigned to the whole group, with particular pin-tryouts for each person. The repetitive nature of this job of trying out particular combinations of digits can be compared to the often hidden (or invisible) crowdsourced online microwork or the low-paid digital piecework of data labeling and editing. These kinds of digital labor are tried out by a vast number of Indian youths who have only problematic employment options, which include working in a scam call center, in the face of rising unemployment.[59]

Can this type of scam work be compared to the above-mentioned kinds of hidden digital labors that have been described by some as "ghost labor"[60] or the commercial content moderation happening behind screens in the "shadows of social media"?[61]

While in the case of the show *Jamtara*, the ethical dilemmas of involving political party workers and school children in scam work is somewhat glossed over as participating in a noble cause to defeat the villain Brajesh Bhan, the stakes are somewhat different in many instances where such scam work is coerced and operates at a very large scale. During the COVID-19 pandemic, criminal syndicates ran industrial-scale scam centers in Southeast Asia, where scammers sophisticatedly enticed romantic relationships and then directed their victims into repeatedly investing in cryptocurrency schemes. Initial successes in these ventures lead the victims to invest more, only to find themselves embroiled in a maze of audits, taxes, and fees, with no way to extricate themselves but to keep on paying. This form of scam has been referred to as *shāzhūpán* (pig butchering scams) because targets are "fattened" before they are "butchered."[62]

The irony of such terrible scam stories is that the sophisticated scammers have themselves been trafficked from Vietnam, China, and Taiwan into Cambodia, Myanmar, and Laos and subjected to physical violence, thrown into debt, and coerced to undertake scam work.[63] In such scamming operations, we see the convergence of uneven mobilities of humans, data, and money. Sophisticated scammers find themselves captured in scamming compounds and forced to work on various scams from online shopping scams to romantic-cryptocurrency scams to online gambling scams. While Sihanoukville in Cambodia has been widely discussed as the epicenter of such scam factories, the lax regulatory laws in the country have resulted in scamming compounds mushrooming throughout Cambodia in "casinos, hotels, resorts, residential developments and office complexes" with "bars on windows and balconies and barbed wire fortifying surrounding fences."[64] The scale of such cyber-fraud, in terms of both the money and labor involved, is astounding. This phenomenal scale of scam operations suggests the need to discuss scam labor in relation to other kinds of (in)visible digital work and the future of work conversations.

Revisiting Scams, Black Money, and Mobile Money

In India, the term "scam" has often been associated with large-scale political or corporate corruption scandals such as the "2G scam," related to illegitimate auctioning of telecom spectrum, and the "Coal Gate scam," where major coal blocks were allocated without competitive bidding.[65] These scams involved large-scale bribes (*ghoos*) of "black money" for "privileged access to public goods" like mining permits and spectrum licenses.[66] The term "black money" is also invoked in discussions of everyday corruption involving small difficult-to-trace bribes paid to low-level state officials for getting an electricity connection or a driver's (vehicle) license as well as to refer to undeclared liquidity in the form of property, nuggets of gold, or piles of discretely stashed cash. The conversation about black money perhaps draws the most public outrage in India when it refers to monetary profits siphoned off to Swiss bank accounts, which Prime Minister Modi vowed to bring back once elected.

What these diverse examples suggest about black money is that it can be something *fixed*, kept away at a secure and secret place (gold in lockers, property, cash, offshore bank accounts), but also something that circulates, and hence is *mobile*, as part of illicit transactions (cash bribes, suitcases full of cash exchanged or passed along, or money moved to accounts of the Swiss bank).[67] Similarly, digital/mobile money also can remain in a difficult-to-trace e-wallet or be repeatedly routed across ATM cards, payment links and gateways, and e-wallets. So, beyond the rhetoric of black money, even after demonetization and the consequent digitization (read: formalization) of money, the storage and circulation of illicit money continues in both analog and digital forms, as is depicted in the second season of *Jamtara*, where Sunny and Gudiya keep competing with Brajesh Bhan to raise more and more money to fund their respective election campaigns.

With the demonetization move in 2016, Modi also played to the tune of the World Bank and International Monetary Fund, which have often identified corruption in southern/"Third-World"/developing countries. These global financial institutions have suggested to developing countries like India that for efficient and transparent governance, corruption should be weeded out by formalizing the economy.[68] This justification purportedly is premised on the argument that with digital money payments, more monetary transactions than before will leave a digital trace, and hence corruption will be regulated. While the Netflix show *Jamtara* does connect the money acquired from mobile phishing to election spending and political corruption, the mobile phishing scams discussed in this chapter are not exactly large political or corporate scams (with the exception, to some extent, of the industrial-level pig-butchering scams undertaken in the scam compounds of Cambodia). Rather, they are financial consumer scams where an individual or one private party decides to con or fraud another person/private party who is pulled into the interaction on the false promise of a reward in the form of tangible/intangible goods and/or financial services/returns.[69]

Scams indeed test regulatory limits. The phishing scams discussed in this chapter are pushing against the limits of financial regulatory protocols of banks, online social awareness of netizens, and surveillance capabilities of cybersecurity systems. At the same time, by revealing the vulnerabilities of these socio-technical systems, scams are reshaping the norms of financial communication systems. Financial scams are more than just aberrations or exceptions to contemporary capitalism. The scams perform critical boundary work to lay bare the workings of financial infrastructures in places that circulate money and deliver financial services at a moment when consumers/users are increasingly conditioned to carry out transactions online over smartphones.

By discussing how phishing scams are challenging the logistical limits of tracing phone calls or evading the data trails available to cybersecurity experts, this chapter adds to the constellation of possibilities and limitations that arise as part of emerging mobile media distribution assemblages, something that I have tried to encapsulate in the term mobilities (un)limited. The characters Sunny and Gudiya in Netflix's *Jamtara* take flights through insurgent trajectories but, even after amassing money from phishing operations, finally find themselves trapped in India's entrenched caste and gender norms. Mobilities (un)limited is not just about the technical logistics of mobile media distribution but is also about the complicated pathways of aspirational socioeconomic mobilities. Access to credit can provide a ladder to neo-middle class populations toward their aspirational goals. A loan can help with paying their children's school fees or allow them to buy tools for their professional work or start a new enterprise, but such loans associated at times with hidden interest rates if not paid back, can also lead to indebtedness. In the next chapter, we look at another financial scam, the microloan app scam, and the argument there is that if we study such scams as constitutive of financial capitalism, we will understand the intricacies of mobile money distribution systems and financial technologies (fintech) operating today.

5 Lending: Loan App Scams and Fintech Services

Digital payments and lending had become almost routine in Indian cities when I was doing my fieldwork in the summer of 2022. In grocery stores, people paid through mobile apps like PhonePe. One could inevitably find a Paytm machine in an open-air bazaar among vegetable sellers and fish being cut on standing blade *botis*. People were also taking out loans in fifteen minutes on their phones, mentioning to me how excited they were about the quick approval process.

At the same time, I heard horror stories of online fraud and how one wrong click could lead to parting with financial information that could cost people their life savings. The government and corporations wanted all Indian citizens to participate in digital money initiatives and, at the same time, cautioned them to be careful about sharing money-related information over the (mobile) internet. In August 2022, I came across prominent ads in Mumbai and Delhi that, when juxtaposed, summed up people's ambivalence about internet finance and loan apps in India. While heading toward the Mumbai airport at Andheri, I saw an LED display touting how easy it is now to receive personal and MSME (micro, small, and medium enterprise) loans: one could get a loan within fifty-nine minutes (see figure 5.1).

The instantaneity of the onboarding process was attractive, with verifications and checks all efficiently handled in under an hour. The same week in Delhi, I noticed another ad, this time a public service announcement from Google, the Ministry of Information and Electronics Technology, and the Digital India initiative. This ad, on a large billboard in the Defense Colony area of Delhi, had veteran Bollywood actress Neena Gupta with two fingers held in a victory sign caution about two things: "Never click an unknown link" and "Never share any bank details" (see figure 5.2).

Figure 5.1
Instant loan ads by Mumbai Airport. Photo taken by author.

Figure 5.2
Ad recommending not to share bank details. Photo taken by author.

Since 2020, I had been researching predatory microloan (instant loan) apps in India, which onboarded someone in fifteen minutes and, within those fifteen minutes, collected all kinds of data from the user's phone. These rogue loan apps carried misleading advertisements on YouTube with attractive (albeit distorted) interest rates and repayment windows. A click took interested borrowers to the Google Play store to download the loan app. Could one really participate in the mobile money transition without being cheated at some point? How would one know when to share bank account information and when not to? If the loan process was going to be instantaneous, how could one hesitate before clicking a link or sharing financial information? The convenience of instant loans seemed to be mixed with the (inconvenient) anxiety of possibly being taken for a ride by these loan apps.

At the onset, I will admit that this chapter has a lot of ground to cover. This is because it refuses to treat a critical current issue like the predatory loan app scams as an isolated case study. It analyzes this issue in relation to the (legitimate) lending ecosystem as it exists in India today, which has been built on venture capital investments and the alluring story of financial inclusion and has a history in rural microfinance and then in telecom policy. Therefore, it becomes important to discuss, as I do in this chapter, the longer historical debate on mobile money and social development.

In the last chapter, we noted the concomitant rise of cell phones and digital money and how that made mobile payments to digital wallets easy and very much a part of everyday life. Since then, the convenience of mobile payments and digital money has moved to areas of digital lending (or platform lending), as well as digital insurance. The embrace of digital money was further accelerated when, in 2016, the ruling BJP regime devalued/withdrew existing Rs 500 and Rs 1,000 notes in a so-called demonetization drive/shocker. This is something that both the banking sector and finance and technology (fintech) companies had been looking forward to and lobbying for, as it allowed the fintech companies to reach more customers and deliver financial services to them on their phones.

Some see this as an opportunity to serve vast swaths of marginalized Indian customers who were previously excluded from formal banking systems but now can be included because they have a mobile phone connected to the Aadhaar identification system (also known as the unique identification system). Others see this rise of digital/mobile money in India as consistent with a wider global trend, where "digitization of payments enables the digitization of finance more broadly," with partnerships between Big Tech and Big Finance emerging as fintech and championing a cashless economy and society.[1]

Fintech companies in India tend to be venture capital–backed startups, as is the case also in other emerging economies like South Africa.[2] The melding of mobile phones and fintech at a significant scale happened in Kenya with telecom company Safaricom's mobile phone–based payment service, M-PESA. Since then, additional lending services were built on the original infrastructure that integrates "mobile telephony and digital data with commercial lending."[3]

Interlinked digital platforms were enabled by the Aadhaar system, the Unified Payments Interface (UPI), and the India Stack infrastructure, which allowed for smooth transfers across bank accounts. The government and the fintech companies declared that "unbanked" Indians could begin to have a credit history and be integrated into digital lending and borrowing systems. Credit distribution no longer remained restricted to members of the formalized workforce who had bank accounts, PAN cards, and credit history, but now could potentially be democratized to include small businesses, the urban and rural poor, and students. From 2015 to 2019, as digital payments became more frictionless, instantaneous, and ubiquitous, digital lending grew as well. Some commentators have welcomed this increased spread of digital lending in the name of "financial inclusion," while others suggest that it has caused increasing debt among borrowers and monitoring of customers' personal and social data for credit scores.[4]

These developments witnessed a further twist in the early days of the COVID-19 pandemic. In April and May 2020, a new variant of digital lending apps (like UdhaarLoan, CashBean, RupeeFactory, and Wifi Cash) mushroomed in India, providing swift microloans to the poor and unemployed, as well to those who were desperate for money because they had lost their daily jobs due to lockdowns. These apps required quick repayments with high interest rates. Unlike banks or financial institutions—which demand stacks of paperwork and substantial collateral and which take many months to approve a loan—these quick microloan apps approved payments in a few minutes. There was a catch. Exploiting the desperate poor and unemployed people who were battling the crisis of the pandemic, these predatory loan apps were found to be flagrantly flouting financial regulatory laws and consumer protection guidelines.[5] They made their debtors/customers install the app on their smartphone and share their mobile's GPS location, contact list, and picture gallery.[6] The collateral for easy loans was user data. Or, put another way, digital lending was seen as a trade-off between the "platformed convenience" of financial activities and concerns about privacy/surveillance.[7] In October 2022, when I tried using some of these loan apps, such as TryCash and FlashRupee (both have since been taken down) on the Apple and Google stores, I was prompted for a government ID as well as a video selfie, which included instructions for how to take the selfie. After that I was asked for a range of device permissions including contacts, call history, and SMSs, and I could not proceed to the next steps of loan onboarding unless I consented to those permissions.

Some of these predatory loan apps do not stop there and harass customers who are unable to pay on time. The term of the loan is short (often seven to fifteen days), and inability to pay leads to abusive warnings and extractive calls and text messages from tele-callers working for these loan apps. Soon after, the app's operators use their access to the borrower's contact list to create WhatsApp groups of friends and family, in which they brand the user as a "fraudster" and threaten them with fake court notices and police complaints. These predatory loan apps therefore go well beyond just being "economic" payment technologies; they set in motion a "communicational politics of circulation."[8] The apps decide who has money and who doesn't, and also what kind of data to collect and to whom they should publicize information about the debtor's financial failures. Clearly, questions of privacy and publicity, of shaming and defaming, emerge as the moral contours of debt and credit are redrawn.

The loan apps at once provide a sense of privacy for people who can now take out loans in the anonymity of their home without having to go out and face the bank officials or their local *sahukar* (moneylender). However, they do not know that the loan app collects data from their contacts and that their relatives will eventually know that they took out a loan if they default. These new recalibrations of moral, familial, and social boundaries are part of the mobile-internet-created financial subjectivities that I continue theorizing about with the term "mobilities (un)limited" developed in earlier chapters.

This situation is not something peculiar to India, as other scholars, such as Rao and McDonald, have mentioned how lower-income populations in China have grown increasingly used to taking microloans with the rise of mobile "alternative finance."[9] With the pandemic and lockdown, these debtors had a difficult time making payments, thereby finding themselves in a problematic situation with exploitative recovery agents.[10] Rao and McDonald discuss some of the strategies devised by working-class Chinese defaulters as a response to collection callers.[11] In this chapter, I will also mention some of the measures taken by consumer rights groups and journalists in India to help those in the debt trap and bring to light some of the concerns about online predatory lending.

My focus is not just on the lenders of risky loans and the vulnerable borrowers struggling with payments. Trying to understand the circulation of money, data, information, and gossip around these loan scams and their communication networks, I study the emergent sociality around digital money among an assemblage of actors, organizations, and technologies—including cybersecurity consultants, tele-callers, digital-privacy advocates, data brokers, nonbanking lenders, software development kits (SDKs), machine learning algorithms, alternative credit score agencies, application program interfaces (APIs), start-up funders, internet ethicists, and digital wallet experts.

My interest in studying the communication networks of these predatory loan apps is consistent with the emphasis in recent scholarship about online scams (in science and technology studies as well as media and communication studies). This research focuses on the networked aspect of such scams and the need to attend to the "role of institutional dynamics and collective behaviors" while examining them.[12] Rather than fitting into an individualized framework that looks at one mastermind con artist, online scams are understood as "networked, collective activity."[13]

As with other scams, the predatory microloan apps had an element of stealth and obfuscation in the way they were putting surveillance malware in the mobile phones of new customers who were downloading their app and agreeing to the problematic terms and conditions. In a general view, one could say that scams and fraud are embedded in the digital economy. I would further add that, while the predatory loan apps were far more intrusive than the usual loan apps and openly flouted rules of online lending, they cannot simply be dismissed as illicit and hence outside the formal online lending process. These predatory loan apps were working with a particular kind of surveillance technology that even some legitimate lenders have used for preparing alternative credit scores. The predatory apps have not just appeared out of nowhere: The financial technology to scrape the data they rely on has been developed by data brokers since 2010. So, the amount of shock expressed in news outlets about the illegitimate surveillance practices of such illicit loan apps seemed a bit excessive, given that very few people in India show any concern about the way social media data are used by legitimate lenders for building credit scores for the previously unbanked who have traditionally not had any credit history.[14] I am not suggesting any easy equations between legitimate and illegitimate loan apps but am only pointing out some of the similarities in the way the data and money trade-off is being negotiated by both.

What is crucial to take into account is that loan apps are situated within particular intersections of financial corporations and software technology firms.[15] Following the platform business model of multisided markets, these loan apps intermediate between users (customers/debtors) and service providers (which include, for the most part, nonbanking lenders).[16] These loan apps, like other apps, include software packages. When particular inscribed libraries, such as SDKs, are analyzed, one can find how other stakeholders—like alternative credit scorers and identity-verification checkers—who supply these financial services, also called fintech as a service (FaaS), are embedded in the lending system.[17]

The nonbanking lender—in India, often called the nonbank financial company (NBFC)—that collaborates with the loan app may not even know whom the app is partnering with regarding a specific service. The invisibility of some of these FaaS firms leads to a partial understanding of how loan apps work, and as I argue in this chapter, this creates an incomplete comprehension of how predatory loan apps have been able to exploit fintech infrastructures and regulatory loopholes. In this chapter, borrowing from the work of Dieter et al.'s "multi-situated app studies" research approach, I situate loan apps within multiple infrastructural settings, including the app store (specifically Google Play store), as well as the financial infrastructures operated by credit scorers and identity verifiers.[18]

In what follows, I will begin with some of the historical context regarding how the landscape of mobile payments and digital lending developed in India, including its relation to an earlier moment of microfinance initiatives. Drawing on some of my research interviews, I will next elaborate on the specifics of the predatory loan app scams, the way customers and regulatory agencies responded to them, and how the actors carrying out the scams also reconfigured their collective networks based on responses from the Reserve Bank of India (RBI), Google Play store, and customers. Following this, I will more specifically ascertain how the regulatory loopholes—both at the governmental level with RBI and at the platform level with the Google Play store—were exploited by these fraud loan apps. Are the rules for mobile payments and online lending a bit lax in India because the government and financial corporations want to promote such lending practices? What steps have RBI and Google taken since these loan app scams were brought to their attention in April 2020?

I will end with a discussion about the general online lending environment in India and why the government and fintech corporations believe that increased digital money circulation, supported by a mobile payment and identity authentication infrastructure (India Stack), is the path to progress. In such a financial environment, the widest possible range of people, along with their money and data, are brought into the fold of capitalist firms, leading to, as some suggest, the "financialization of daily life."[19]Another perspective would be to argue that this financialization has not suddenly happened now. Rather, everyday domestic and personal life has always been entangled with financial calculations, but there are indeed noticeable shifts in financial subjectivities as state-corporate bodies push fintech products into the intimate domains of their consumer-citizens' lives and their kinship networks. This chapter studies emerging financial media forms and technologies such as account aggregators and loan apps that are intensifying financial calculations and speculations in everyday life: the rise of such credit infrastructures is taking place along side a moment of heightened aspirations which may result in encouraging extensive borrowing among Indian citizens.[20]

Digital Lending in India

The story of digital or mobile lending in India has to begin with the story of digital/mobile payments. Then I will go back to how the narrative of microfinance in rural South Asia has been transformed into one of digitally catalyzed financial inclusion, with a stress on how digital data supposedly offer a more complete picture of a poor person.

On November 8, 2016, the government of India announced the demonetization of all Rs 500 and Rs 1,000 banknotes. Many observers suggest this moment as the catalyst for India becoming a place for digital money, as now a significant portion of the Indian population was compelled to shift to digital payment services like Paytm in the face of a paper cash crunch. Paytm was introduced in 2010 and initially allowed users to pay digitally for selected services like mobile recharging and shopping bills.[21] Some youngsters I interviewed recalled that they had heard about Paytm long before demonetization, and some had even used it, but their parents, until demonetization, had not considered digital payments. Some consumer rights organization members I chatted with opined that, with demonetization and the promotion of digital money, cash was privatized and cash issuance—which was once the sovereign function of RBI, under the direction of the government of India—was now to a significant extent dispersed into the hands of market forces. Such consumer rights organizations likened this privatization of cash to other moments of privatization in India, such as in the telecom and aviation sectors after the neoliberal reforms of the 1990s. Demonetization might have sent ripples across the money-circulation infrastructure in India, but indications of a change toward a "less cash" society were provided a bit earlier as well.

A member of a consumer rights group who has been tracking the story of mobile money in India told me that, before demonetization, it was the launch of the Unified Payments Interface in April 2016 that made him take notice. UPI, created by the National Payments Corporation of India (NPCI), is an interface for financial transactions that consolidates multiple bank accounts into a single mobile phone application, thereby creating a digital payment gateway. This open network for payments meant, according to this consumer rights advocate, that now the banking world in India was entering a more tech-friendly space. The soft launch for UPI was on April 11, 2016, with twenty-one participating banks; it was more publicly launched in August 2016. The press release by NPCI on UPI's soft launch already declared, "We are confident that several banks will join UPI this year and the number will multiply further. Our focus is in line with RBI's vision of migrating towards a 'less-cash' and more digital society."[22]

UPI is one of the many application program interfaces in the India Stack. This set of open/public APIs—including Aadhaar-related authentication and electronic know your customer (eKYC) documents—allows for, in the words of India Stack creator iSPIRT, "ease in the integration of mobile applications with the data securely stored and provided by the government to authenticated Apps."[23] Thus, Aadhaar and UPI are (kind of) designed for credit, because these APIs generate data that can be used for an alternate data-based credit score, aiding financial services, through India Stack, to target the millions of consumer-citizens of the second most populous country in the world. As ISPIRT notes, "The India Stack enables Apps that could open up many opportunities in the financial services, healthcare and education sectors of the Indian economy." In fact, a fintech journalist I met in Mumbai said that proponents of India Stack have the same pitch as Muhammad Yunus's Grameen Bank and the microfinance initiatives: Access to credit is considered the golden mantra for poverty alleviation.

Microfinance—the idea of advancing microloans (having ticket-sizes of just a few hundred or thousand rupees), initially often for poor, rural women (as in Yunus's Grameen Bank initiative)—is viewed by many as a way to solve the problem of poverty. Microfinance could also be seen as a way to integrate the poor into circulations of credit and capital.[24] This notion of incorporating the bottom billions into the formal, capitalistic economy is based on the ideas of economist C. K. Prahalad, who asked management thinkers and development economists to carefully attend to the sociocultural practices and business models of people at the bottom of the pyramid as a way to create new profit-making opportunities for corporations.[25] In this explosion of "bottom of the pyramid capitalism," Sohini Kar remarks that "the poor are no longer considered just passive objects of state-led development but active market participants as consumers and entrepreneurs themselves."[26]

Whether the poor have really benefited from such microfinance initiatives remains ambiguous at best, but microfinance initiatives have now been further transformed by new values attached to algorithmic data, smartphones, and what such technologies can bring to financial inclusion.[27] Steve Hollingworth of the Grameen Foundation, a nongovernmental organization that works to replicate the Grameen Bank model around the world through a network of partner microfinance institutions, notes that "mobile money, as a channel, offers an opportunity to reach poor clients quickly. It took microfinance 30 to 35 years to approach 200 million clients, whereas mobile money platforms are expanding more rapidly and provide a big opportunity." If the mobile channel leads to faster outreach than the face-to-face approach of earlier microfinance initiatives, the digital data gathered, according to Hollingworth, is the "lever" to combat poverty, helping to "design solutions" for the specific needs of the poor.[28]

The India Stack—as the largest set of APIs, with applications for mobile phone data authentication and aggregation—allowed for the kind of interaction between government, citizens, and fintech players that the Grameen Foundation was looking for. The premise of India Stack APIs was that you could now extend credit to a wider population, many of whom were not part of the formalized workforce and did not have credit cards and PAN cards (issued by the Income Tax Department of India). India Stack made it easy for a lot of Indian citizens to sign up for digital payments and lending apps.

But easy onboarding is one thing, and persisting with an app in the face of scams is another. With increasing UPI transactions, the risk of fraud has also increased. Fraudsters sometimes monitor mobile phone screens, and customers have been exposed to unauthorized payment links and sham UPI handles.[29] Aadhaar data have proliferated in the black market, and fingerprint cloning has become rampant, so Aadhaar-based payment frauds are also flagrantly spreading. Biometric markers like fingerprints and iris scans are considered to provide uniqueness to a person's identification process carried out through Aadhaar, but fraudsters are cloning fingerprints from digitized land deeds and other government records. The equipment, such as epoxy putty or candle wax, that is used to copy a physical fingerprint from a scanner (or a surface where the person has left a mark) is easily available.[30]

Continuing with the story of digital lending, it is important to look at the first constituencies targeted by fintech companies. When it went live in December 2016, mPokket became one of the first successful microloan apps meant for students who needed urgent cash. Most students in India were ineligible for credit cards, and they were likely to face cash crunches because they were too young to have saved anything substantial. For the small loans they required, students were just not worth the time and resources of banks and other traditional financial institutions. mPokket simplified the sign-up process for students: They could find the microloan app in the Google Play store, download it, register through their Google or Facebook account, and then upload eKYC documents. Once verified (in most cases, within an hour), they became eligible for loans. The ticket size for these loans was small, from Rs 500 to Rs 20,000. Timely loan repayments would help increase the borrowing limit for the next iteration.[31] mPokket is an example of a small, focused digital lending experiment that has evolved with improvements in digital verification and payment infrastructures such that it has become easy and quick to disburse loans and recover them. It goes without saying that the digital lending industry has been very focused on mobile phones as the primary channel, and students—being the youthful, savvy users of smartphones—were a readily accessible market for these microloan apps.

Once consumers started using smartphones for borrowing, the question for lenders became: How do you gather alternative data? The alternative data collection and scoring practices on which creditworthiness is evaluated remain opaque. After all, mPokket—or, for that matter, any microloan app—is not simply basing the borrowing limit on prior repayment history. Comprehending how phone surveillance and data collection practices by data brokers have changed since 2015 is key to understanding credit ratings and digital fintech processes. This is where a whole ecosystem of players like alternative risk-score providers come into the picture. These companies collect data both directly and indirectly, sometimes even before somebody has installed an app on their phone. Describing the situation in the mid-2010s, Srikanth Lakshmanan from the consumer rights organization Cashless Consumer notes, "You may just say [you are] interested in an app—you provide a mobile number, and the loan provider will send you an SMS to download the app. What you do not know is, by that time the loan provider has [an] alternative data score that he has managed to get from one of these data brokers, because your data was already there."[32]

And how was the data already there? Lakshmanan notes that the Android permission system was pretty weak prior to Android 8 in 2017. SMS messages were programmatically accessible in the Android permission ecosystem, and furthermore, after you installed the microloan app there was no way of revoking particular permissions. In 2017, this aspect changed when Android 8 introduced more granular permissions. Between 2015 and 2017, data brokers could scrape data using a gaming app or an mp3 music app in which tracking software was embedded, lifting users' SMS texts. Gopal Sathe wrote in a *HuffPost* article about CreditVidya, a Hyderabad-based alternative credit-rating company that ran its snooping code in many seemingly innocuous music-related apps.[33]

Prior to Jio phones and their data plans, streaming on YouTube was not as widespread, especially among people who did not use (or have) a computer with stable bandwidth. At that time, for users who used a mobile phone as their only digital device, downloading mp3 software and songs through apps was a popular practice, and CreditVidya used it to their advantage. Lakshmanan explains the steps:

> If you want to listen to music, you would actually have [an] mp3 on your phone, and most of us have an mp3 on computer and then transfer to phone, but there is a section of population whose only access to computing resource was their smartphone. . . . So there was this concept of mp3s bundled into an app. So you could go to a [Google] Play store and then, say, type R. D. Burman hits, and there will be an app which has links to songs of R. D. Burman—mp3 songs inside an app, and you would install that app. This particular provider actually bundled a spyware,

> which essentially collected all the SMS history, location history, and so on and so forth . . . and they became an alternative data provider to CreditVidya, which is one of India's alternate data-score provider[s].[34]

In 2017, the music-streaming app JioSaavn (from Jio) and Prime Music from Amazon arrived, both of which could be used with the 1 GB of data per day offered by Jio and other telecom services. As a result, the need to download mp3s of Ilaiyaraaja or R. D. Burman hits was reduced. Google also responded to global outrage against these privacy lapses and emerged with Android 8, which was stricter with app permissions and more cautious regarding background threats.

I have discussed these CreditVidya practices to emphasize that the predatory loan apps that proliferated during the COVID-19 pandemic were not the first to commit snooping operations on customers. In fact, some privacy experts have argued that the data brokering industry has existed in India since the early days of the internet when leaked employee data from an online job portal like Naukri.com could be found circulating on CDs. In too many cases of financial scams or job scams or entrance exam question paper leak scams, Snigdha Poonam and Samarth Bansal argue, the government authorities go after scammers rather than the data brokers who provide them the material trawled from data storage websites or extracted from bank officials by cultivating relationships.[35]

The rise of these predatory loan apps was met with outrage in the media (like Zee Business's "Hafta Vasooli" campaign), but this coverage did not mention the history of how personal data was often used as collateral by loan apps in the past. The way the predatory loan apps used the data they collected to send WhatsApp messages to the borrowers' contacts may not be how data is used by alternative credit scoring firms and their clients, but the collection of so much private user data—including phone browsing history, GPS information, and social media accounts, which can potentially be leaked or used by nefarious interest groups—should by itself be a cause of concern.

Alternative Data for Assessing New-to-Credit Customers

CreditVidya's use of tracking software in seemingly innocuous apps is brushed aside by the company's advocates, who maintain that it is an early mover and innovator in the credit risk assessment space. CreditVidya collects lots of data (500 GB every day) and has developed proprietary algorithms to build credit scores, taking into account ten thousand data parameters related to the socioeconomic profiles.[36] Kailash Nath, a senior associate at Bharat Innovation Fund, which has invested in CreditVidya, notes that CreditVidya's efficacy and credibility is built on big data analytics, psychometrics, and its ability to instantaneously automate the credit-underwriting process, thereby making the transaction easier for both lenders, like microloan apps, and borrowers, such as new-to-credit customers. All that the digital lenders have to do is plug CreditVidya's software development kit into their app, and CreditVidya's built-in algorithms do the work of ascertaining creditworthiness.

Another key player in the Indian fintech space, and specifically with regard to credit scoring and identity verification, is Lenddo. Lenddo was an early entrant into the fintech system in 2011, offering loans and gathering data in the Philippines and then in Colombia and Mexico. Using small loans ranging from $100 to $2,000, in its initial years Lenddo was able to test its algorithms for credit scoring. Soon after, it stopped offering loan products and positioned itself as a credit-rating company. Thereafter, Lenddo sold its credit-scoring and identity-verification services to other banks and loan apps, and its products have been integrated into the onboarding process of digital lenders.[37] Lenddo entered India in July 2016, and the company distinguished itself from other financial players—who were still reliant on credit bureaus—by using a mix of social media digital footprints and psychometric data pulled from application forms, even as some doubted the efficacy of this method in the Indian context.[38]

To make its predictions, Lenddo's machine learning algorithm is trained on borrowers' social media data, particularly Facebook and LinkedIn. As the company's founder and CEO Jeff Stewart noted, "Lenddo's credit scoring algorithms are built on top of the massive amounts of opt-in alternative data such as Facebook and LinkedIn."[39] Marie Langevin interprets this to mean that Lenddo links credit scores to the individual debtor's "capacity to form online groups with friends, parents and colleagues."[40] For Lenddo, the social graphs of the debtors' online relationships reveal character traits such as reliability, responsibility, and honesty. Langevin thereby argues that Lenddo privileges artificial intelligence and behavioral economics in evaluating creditworthiness instead of "drawing on the know-how concerning the evaluation and measurement of the productive capacities of debtors"[41]

A careful evaluation of the productive capacities of debtors would require loan apps to (perhaps) deploy in-person, human expertise and invest considerable time: credit analysts would maintain a close relationship with microentrepreneurs to assess cash flow and repayment potential. But time and human labor are precisely what the alternative credit scorers are promising to cut back on compared to earlier microfinance initiatives.

At best, these alternative credit scores—partly relying on social media data for character and behavioral traits—can provide some indication of whether a particular debtor can be relied on to pay on time. However, if such a debtor simply has no means to pay, how will they? This leads Langevin to argue that "given that these algorithms do not measure productive capacity, but rather the social or character traits that indicate a propensity to repay, they can lead to a situation where credit is extended to those who cannot hope to repay and only contribute to their further impoverishment."[42]

Beyond just Lenddo, this is a discussion about data-driven algorithms supported by increasing computational power and machine learning techniques not only telling stories about people but also steering consequential decision-making about them, including whether they are eligible for loans and of what ticket sizes. Using social media connections or phone contact data as alternative credit scoring data can have benefits but also be compromised in contexts such as India, where in peri-urban areas, as studies have indicated, there is a possibility that women's phone use and access is mediated by a male relative. In such a situation, women feel discouraged from having too many phone contacts and self-censor their text messages and social media connections.[43] This would lead to a thin data footprint capture or erroneous/incomplete data gathering for the algorithms to make decisions. Since smartphone ownership and usage among women in India is much lower than that of men, this kind of alternative data gathering mechanism has limitations and can perpetuate exclusion (rather than inclusion) of women in financial activities.

The predatory loan apps, about which I will elaborate in much greater detail in the next section, were interested in catching debtors in a debt trap. These alternative credit score generators might, unwittingly, just on the basis of their errant algorithms, be doing the same. At the point of onboarding to a legitimate loan app, debtors are asked for opt-in permissions by Lenddo SDK, which scrapes information about their digital footprint and browsing history. That said, the moment when all the user permissions are sought is also when the debtor is made to feel that the onboarding process is seamless and effortless, thanks to speedy verification procedures and rapid payment infrastructures. This serves to obfuscate the complexity of the multiple actors, intermediaries, and software applications involved in data accumulation and analysis. This rather fleeting, instantaneous moment involving a couple of clicks and some scrolls is never quite enough for customers/debtors to comprehend what data is being monitored and how it is going to be used and by whom for what purposes. These questions will keep appearing as we discuss the particulars of predatory microloan apps and the tactics they deployed during the pandemic.

Deceptive Lending Practices

It was no doubt opportunistic of the rogue loan apps to spread their tentacles in India just as the COVID-19 pandemic struck in March 2020. With no jobs and a strict lockdown, people were desperate for short-term (and payday) loans, hoping to pay them back as soon as they received their next paycheck. These illegitimate loan apps levied hefty interest rates, which were further increased with additional processing fees and charges.[44] Salaries kept getting deferred, and that affected debtors' capacity to repay. These illegitimate loan apps then upped the ante by having telecenter-based recovery agents make threatening calls to the borrowers, listing penalties they will have to pay.

The predatory loan apps used the customer data collected during the initial onboarding process to harass those who had defaulted. Intimidation practices included calling not just the defaulter—and threatening to sue them and block their PAN card—but also the defaulter's friends and relatives (from the contact data captured earlier) and using expletives against them. So blatantly vulgar was the language, tone, and demeanor of these calls that some harassed debtors started recording them. These calls were replayed on prime-time news shows, especially those that were part of the Hafta Vasooli media campaign of the Zee Business channel.[45] *Hafta* in colloquial Hindi means the "protection money" that gangsters often ask for. The behavior of recovery agents during the calls they made and the WhatsApp messages they wrote was so egregious that the anchor of a Zee Business show felt they should be compared to the mafia. The first Hafta Vasooli broadcast, in April 2020, converted a prime-time news show into dramatic theater, using soap opera aesthetics to sensationalize the actions of predatory loan apps. To some extent, the show tried to empathize with the debtors in a very difficult moment of their life, and it did bring some amount of public consciousness about the deceptive practices of loan apps (see figure 5.3).

While public outrage and the spectacle of a TV show might have its place, the broadcast did not trace the ways the fraudulent loan apps were using existing fintech infrastructure. The TV show failed to acknowledge that one reason these predatory loan apps were succeeding had much to do with how certain protocols of loan disbursement and onboarding (like asking for opt-in permissions for certain phone data) had been normalized by legitimate loan apps. Furthermore, the activities of these fraudulent firms kept shifting over time, as they tried to find some way to prevent regulators from tracing them. The authorities seemed to be facing a Whac-a-Mole situation: If some fraudulent loan apps were eventually taken off the Google Play store or the police and the enforcement directorate were finally able to trace a bank account connected to a rogue app, other new sham small-term loan apps emerged with novel ways to route their money.

When I interviewed Pravin Kalaiselvan, chairman of SaveThem India Foundation, an organization set up to help debtors entrapped and harassed by predatory loan apps, he explained that many debtors who contacted

Figure 5.3
Snapshot from *Hafta Vasooli* (Zee Business, Apr 16, 2020).

his organization could not go to their parents or file a complaint with the police because they had difficulty confessing that they had taken loans and failed to repay. The police initially could not understand what the problem with the loan apps was, as they did not have a complete knowledge of the insidious tactics being used to recover the loans.

After some time, according to Kalaiselvan, the loan apps began shifting their strategies, making them far more difficult to get rid of. For example, Kalaiselvan notes that when his foundation and the debtors started recording the threatening calls from loan recovery agents, they would find after they hung up that the recorded file was deleted; the evidence they had sought to gather against the loan company was not there. Furthermore, even after paying off the loans and uninstalling the app, the debtors would still get calls and texts from the loan agents. When Kalaiselvan and his team checked the phones carefully, they found unknown applications hidden without appearing as icons on the phone. In the initial process of installation, the loan app had installed malware that was difficult to get rid of, even after uninstalling.

Kalaiselvan and his organization have been tracking the cases since April 2020. By February 2021, a new money-extraction technique had emerged: photoshopped pictures of debtors in compromising acts.

> We found something very strange. We started receiving complaints that they started changing pictures. . . . They create selfie photograph. . . . They will collect all our data. They have facial recognition algorithms. There is a company in Singapore, some AI company. That company's API got integrated in few Chinese applications. . . . They started capturing our face AI recognition. So, once again, it is [a] threat. . . . They [Recovery Agents] started defaming people in their contacts. Even after payments, they [the debtors] kept on paying amounts. This is February 2021. This was completely carrying on, as we worked with Telangana police, in Hyderabad to more than fifty-three suicide cases in India.[46]

Threatening calls about legal notices and PAN card cancellations no longer seemed to be working, as they had been debunked by news media. Instead, threats of defamatory images were being deployed by these rogue loan app firms and their agents. With suicides rising, the police acted as well. But by the time the police raided the address that a loan app had provided while registering with the Google Play store, the offices had already been vacated or were no longer operational.

Some of these offices ran as call centers for collection agents.[47] Journalist Arti Singh interviewed some of these tele-callers, and they confided that there was immense pressure to achieve high collection targets if they were to receive incentives and bonuses. This is a side of infrastructural tracing where one finds a different dynamic of power relations than when one focuses on the interface. An interface analysis would make these tele-caller collection agents the villains of the loan app fraud saga, as they are the ones making the calls. Their voices are the ones that are reaching out to the customer, the debtor. However, an infrastructural analysis suggests the power of the upper management over these collection agents.

The telecenter operations are just one part of the fintech infrastructure of money and data distribution. In the rest of this chapter, I will focus on two more such infrastructural systems: the app store and fintech as a service.

Appropriating Fintech Infrastructures

In most cases involving both legitimate and illegitimate microloan apps, a nonbanking lender, usually called a nonbank financial company (NBFC), partners with a lending partner or a loan app so that it can reach a wider set of people with loan offerings. Usually, an NBFC works with several lending partners and loan apps, and the loan app in turn depends on various actors to supply SDKs/APIs for loan dispersal, credit scoring, identity-verification checks, payment processing, and loan collection.[48] These services are called fintech as a service, and many loan apps depend on a few niche FaaS companies. The NBFC partner may not even know who the loan app is partnering with for a particular service. This shows how a multisituated study of loan apps is so important to understand software as infrastructure and how apps are situated within varied infrastructural relations, such that the work of FaaS remains almost hidden from the NBFCs.

Scholars working on the relationship between apps and infrastructures stress that apps are not "stand-alone objects" but are involved in data flows of multi-sided platforms and connectively entangled with digital infrastructures at various sites that make their operations possible."[49] This is precisely what we see with these fintech apps. Some of the relationalities of these apps can be found from studying the source code, which provides more information about APIs and SDKs, and this is what Suman Kar and Srikanth Lakshmanan from Cashless Consumer did to uncover the various third-party tools used to access the user phone data and software-based functions of mobile operating systems. SDKs are an essential part of the modular design of the app and can reveal information about app templates and third parties (who embed their SDKs in the app and, at times, surreptitiously collect data). Knowing this information could help in resisting and exposing the obfuscation practices of app companies, including microloan apps.[50]

During the microloan controversy, an NBFC like Inditrade remained unaware of the fact that some of the loan apps it was partnering with were using identity-verification services from Advance AI, a Chinese company working on AI-driven eKYC. Some of these verification methods, as investigated by Cashless Consumer experts Kar and Lakshmanan, entail taking a selfie during loan onboarding, which is then sent back to a server in China. This server tests and checks whether the borrower is a real person, after which the information is returned to the app. These transnational seepages of data raise national concerns, as some of these ID-verification techniques involve liveness detection and facial recognition data, which can then be mirrored in other databases. The issues raised by Lakshmanan ask us to understand the activities of loan apps not only at the individual or company level but also at the level of geopolitics and infrastructural platform services (see figure 5.4).

One of the reasons for the Whac-a-Mole nature of the rogue microloan apps is that they are so easy to put together. These predatory loan apps are white-label-based, meaning they are essentially deployed out of the same templates. If and when one loan app is brought down, the company behind

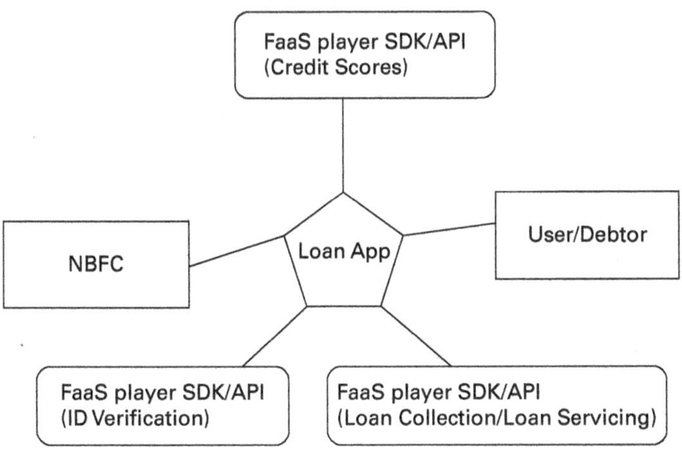

Figure 5.4
Stakeholders/intermediaries plugging their SDKs/APIs for instantaneous loan onboarding experience.

it remains the same, and that company can just partner again with the same set of stakeholders and again approach the three or four white-label providers and then produce an app with a slightly different name and brand with the same stakeholder SDKs/APIs working in the backend. A company like Epoch.api prides itself on being a "one-stop shop for lending apps."[51]

These FaaS players are important because their SDKs and APIs can just be plugged in and the loan app can start functioning. Not every lending app has the time or resources to build all parts of its services, so one way of starting a loan app is to integrate these various SDKs/APIs from different providers. As Arundhati Ramanathan notes, "The use of SDKs and APIs by these FaaS players has allowed the Indian startup ecosystem to be competitive. It allows startups to take on digital behemoths by giving the same level of service."[52] At the same time, these FaaS intermediaries may actually be retaining data that they are supposed to just pass on to the NBFC or the loan app. Tongdun—which provides risk-score SDKs for loan companies like FlyCash, UCash, and KarmaLoan—has been implicated in earlier probes by the Chinese government into illegally collecting and selling personal data.[53]

Not just in the loan app space, but also within the wider mobile app ecosystem, third-party SDKs have become constitutive actors in helping offer monetizing, analytics, and advertising functions that convert user sociality into data at the backend and thereby improve the functionality of the apps, with the hidden collateral of greater surveillance.[54] SDKs are far more integrated into an app's source code than laptop/desktop cookies, which should make us more concerned about user privacy in the mobile phone ecosystem.[55] For loan apps, the implications of permissions requested by third-party SDKs for effective app use are particularly crucial to study given the sensitivity of the information collected. Equally important is the need to ascertain how the app stores are regulating the extent to which these permissions are necessary and reasonable, or appropriate and ethical. With the spread of financial services in emerging economies like India and Ghana, there have been concerns about whether the data generated out of erstwhile unbanked poor customers might be compromised with the rise of customer protection risks. David Medine and Gayathri Murthy, in their CGAP (Cumulative Group to Assist the Poor), report noted that by emphasizing individual consumer consent, data protection regimes end up placing undue burden on low-income populations in emerging economies who have just started opening accounts and going online.[56]

App stores provide users with access to apps based on certain conditions, and they also grant app developers the opportunity to showcase their products so as to maximize hits or downloads.[57] "Logics of optimization" are very much a part of contemporary platform politics, where app developers engineer and design elements in specific ways to make their apps "more searchable, discoverable, usable, and valuable" in the search-engine-dictated visual regime, where thousands of apps are competing for attention.[58] The Google Play store is key, therefore, for the rogue fintech developers to promote, distribute, and monetize their fraudulent microloan apps. Suman Kar notes that such apps focus heavily on app store optimization. They stuff the Android package kit installation files with ten to twelve keywords—such as "udhaar" (Hindi for "loan"), "cash," "dhan" (Hindi for "wealth"), "jaldi" (Hindi for "quick"), and "borrow"—to ensure easy searchability in the app store.[59] Furthermore, several journalists and stakeholders that I spoke to emphasized that these microloan apps were utilizing social media advertising, especially YouTube and to some extent Facebook. As Nishant Kauntia wrote, "fake" ads on YouTube and Facebook contained photoshopped images of celebrities and bank logos endorsing such microloan apps.[60]

These ads had misleading interest rates and distorted payback periods, which enticed a click/tap from the interested borrower. But when the borrower actually landed on the app, they found the interest rate to be much higher and the repayment window to be much shorter. For example, a fraud loan app "CashBus India" advertises using the IDBI bank logo with a 0.25 percent interest rate over twelve months, which is simply not the case. To offer another example, Loan Lucky Cash promoted itself using Prime Minister Narendra Modi's image to suggest that very little documentation was required for getting a zero-interest loan. Many loan apps usually ask for pay stub, Aadhaar card, and bank statement, but this loan app was touting that it required none of these things, clearly a spurious claim (see figure 5.5).

As Google/Alphabet owns both YouTube and the Play store, the company received sharp criticism from consumer rights organizations. They accused Google of being both host and advertiser for these predatory loan apps, controlling significant aspects of the online ecosystem that allowed the flagrant spread of such apps—all without proper regulations, or at least with lax enforcement of prevailing regulatory codes.[61]

Over time, some NBFCs became suspicious about the (shady) activities of the predatory loan apps and wanted to discontinue their lending license.

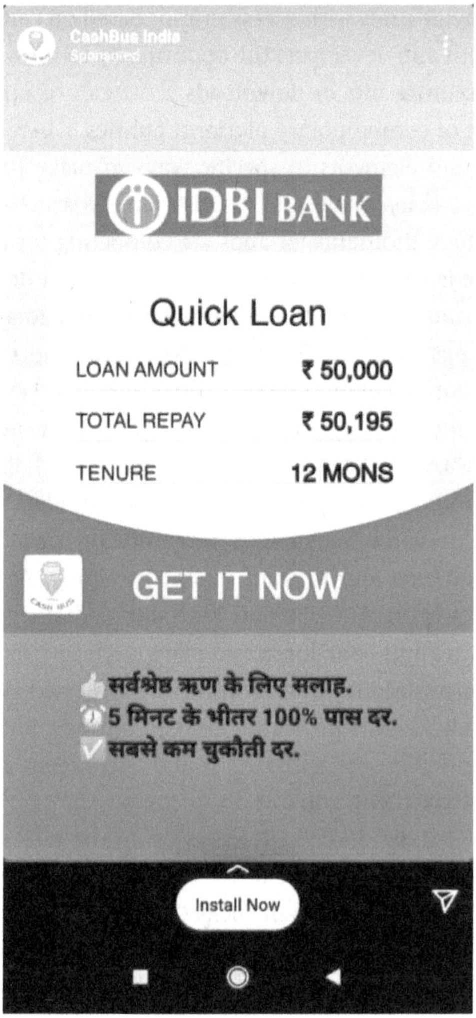

Figure 5.5
Use of IDBI logo and Modi's image in predatory loan apps. Courtesy: SaveThem India Foundation.

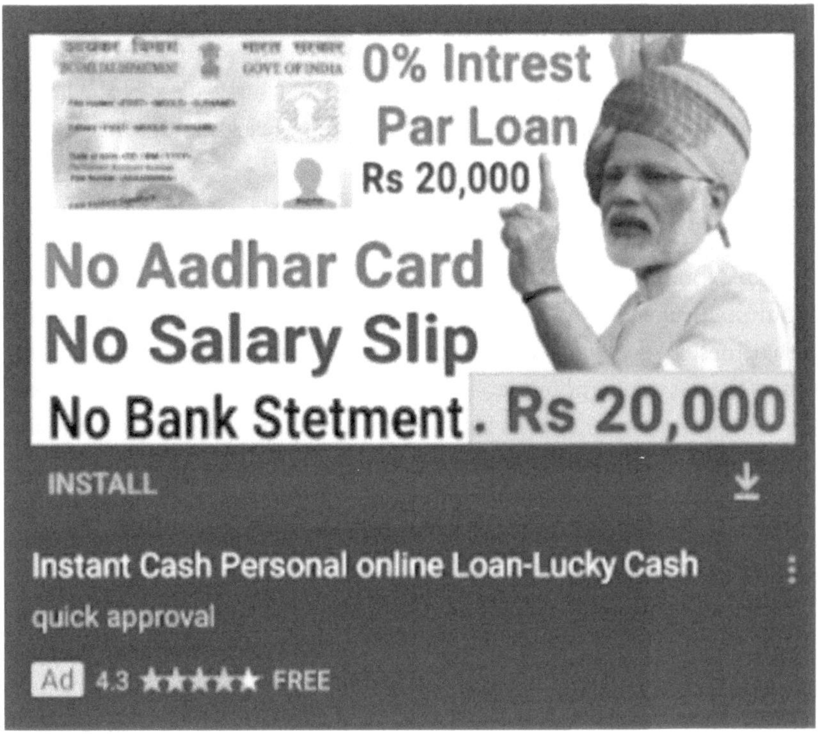

Figure 5.5
(continued)

They were persuaded to stay on board by these predatory loan apps, who wanted to keep using their affiliation with a legitimate lender to further swindle new customers. In order to lure NBFCs, some lending platforms (especially the predatory and illegitimate loan apps) using aggressive tactics assured them that they would completely cover the losses when customers defaulted. This was particularly attractive to NBFCs during the pandemic, when default frequencies were very high.[62]

Following many complaints from victims of predatory loan apps as well as consumer rights groups, the Reserve Bank of India introduced stricter guidelines for digital lending. The RBI issued a circular in September 2, 2022 that laid down several restrictions regarding premature withdrawal of regulated entities like nonbank financial companies (NBFCs) and platforms selling loans on the secondary market. The lending platforms were also cautioned to stop advertising guaranteed returns to attract customers without explicitly mentioning the risky nature of investment.[63] The Google Play store also took steps to take down the illicit apps. The exit of many predatory loan apps created a void in the online lending market, which encouraged established lending platforms like KreditBee and Paytm to aggressively enter this market.[64]

Conclusion

Fintech products are as much about cultural form and affect as they are about political economy and infrastructures, and my point in this chapter is that these two dimensions are connected. These loan companies' promises to the desperate (potential) borrowers as they are lured into downloading the apps, the connections between the ease of digital payments and the consequent growth of digital lending in India, and the way financial capitalism is expanding by including the poorest of the poor in debt traps that are an almost textbook example of "predatory inclusion": all these aspects need to be studied in relation to one another.[65]

At the same time, some stakeholders I conversed with genuinely considered loan apps to be helping to uplift the poor and saving them the disgrace of going to the bank and being humiliated about their lack of land papers, property documents, and credit scores or being tricked by unscrupulous moneylenders. Overall, fintech is not only about selling the instantaneous and convenient experience of loan onboarding but also about making such financial transactions a habit and cultural norm.[66]Fintech data analysts agreed that some alternative datapoints for risk scoring might in themselves seem arbitrary, but they have to be considered in relation to other data points, and some of these data points are not trying to predict the financial behavior of borrowers but ensure their traceability. An oft-repeated example I kept hearing in fintech circles was that of a young person who would call his grandmother every weekend and thus was considered a reliable borrower even though that youngster did not have any dependable funding sources or salary and thus could not be considered creditworthy in the traditional sense. When I put this question to Amit Das, who runs the FaaS company Think360.ai, he told me that "lending" in some cases is "not about giving the money but it is about collecting the money," and this calling your grandmother example suggests a borrower who is close to a person/relative and thus can be traced to recover the money if required through that person/relative: "If I pick the phone and talk to your grandma, she would be able to tell me where you are." So, instead of predicting financial behavior, this data point is a proxy for traditional analog lending practices, approximating "social trust" and the figure of a "guarantor."[67]

Beni Chugh and her team at Dvara Research found that—contrary to conventional wisdom that Indian citizens (many of them new to the internet) are focused on access and convenience and are indifferent about surveillance—"people in India care deeply about their personal data and privacy."[68] In an interview with me, Chugh further elaborated that there are many layers of negotiation regarding the data-credit trade-off and that it is not as simple as somebody taking a user's information and giving money in return. There are different kinds of information—social media data, identity documents, and locational information—that need to be internally differentiated. In order to part with some of this information, Indian citizens, even those who desperately needed money, wanted some guarantees that their data was not going to be illicitly disclosed to third parties:

> Until digital finance, data protection was not a thing in finance. Finance is . . . mostly seen through the trope of confidentiality—tons of RBI papers look at confidentiality . . . nobody really came down to data protection and privacy . . . [Data protection is cast as] trade-off: take their data, give them access to financial services. . . . Sure some data is tradeable . . . some data is expendable. . . . [People are often asked] Are you willing to share some data if that is going to increase your ability to take a loan? We asked this lady if they would share their YouTube history? What if you did not have [a] penny? Children's location. Own location. We discovered a hesitation and a strong sense of privateness in low income community.[69]

Individuals interviewed by Chugh and her research team asserted their right to be assured that their personal information would be handled responsibly, thereby claiming agency over their data. Furthermore, as digital tools open up access to credit for the poor, the calculation of creditworthiness through mobile phone usage is especially problematic for women, who in many cases do not own a smartphone. This makes alternative credit scoring for women based on phone data more difficult and at the same time aggravates the existing problem that women are more likely than men to lack credit history information.

These challenges for fintech are sometimes presented as opportunities. Repeatedly, fintech operations aimed at the poor reference the Grameen Bank, which did target rural women's cooperatives. So powerful is the Grameen Bank story in relation to female empowerment that feminist fintech entrepreneurs I met—who were making sleek new investment products for urban millennial women—kept telling me just how influenced they were by the Grameen Bank model. One of them recounted her aunt's story of having sent her children to study in the United States, all thanks to a women's microfinance initiative in a village in the state of Karnataka. At the same time, as I have repeatedly stressed, fintech loan apps are evaluating creditworthiness based on artificial intelligence and behavioral economics. This is very different from the Grameen Bank model, which is premised on agents actually going into the field and meeting women; assessing their daily and monthly cash flow; accounting for how much land, cattle, carpentry tools, or domestic utensils they owned; understanding their experience as an agriculturalist or vegetable seller; and evaluating their social ties within the women's cooperative.[70] The different contexts in which the Grameen experiment is being evoked, deployed, and spread by fintech companies speaks to the expanding tendency of financial capitalism.

I have traced this history of microfinance initiatives in India because it goes a long way toward setting the tone and perception about loan apps. Some loan apps such as mPokket and Mahila Money have certainly worked out well for lenders and borrowers, helping aspirational Indian citizens on their way to achieving success. That said, I have argued throughout this chapter that predatory microloan apps targeting desperate borrowers should not be dismissed as an aberration in the digital loan ecosystem, because that would be tantamount to a strategic disengagement from the systemic problems with online lending and FaaS infrastructures that makes it so easy for third-party developers to siphon off user data.

Within the Indian context, it is key to point out how important a role the state has played in fostering an ecosystem for digital payments and lending. The Indian government has actively promoted platform lending. An understanding of the storage and distribution technology of India Stack—so closely connected to the mobile devices of India's population—is crucial because data capitalism in India is not just being run by Big Tech but gets its framework and APIs for digital ID and payment gateways from the Indian state. These elements of India Stack flow through and interact with fintech companies in the name of financial inclusion. This is what Jacqueline Hicks terms "digital ID capitalism."[71]

In some cases, India Stack's UPI and eKYC facilities have helped lending platforms engage with underserved communities to shift from last-mile cash delivery to digitized lending; this allows the rural borrowers to get credit scores based on their bank account passbooks as well as through their UPI transactions.[72] That said, the government policies that aim to democratize credit actively encourage many different platforms, not necessarily in the financial sector, to become lenders through the open credit enablement network. This can lead to situations where there is very little regulatory framework for transactions, thereby increasing possibilities for online fraud and compromising data protection.[73]

The movement of data and financial services across the different layers of the India Stack is important to examine as one studies digital lending in India. It is not just Big Tech and finance companies coming together to provide FaaS; the Indian government is also playing an active role in start-ups. "Financial inclusion through platform lending" and the "platformization of financial inclusion" are promoted as key financial infrastructure by a broad coalition of policymakers and businesses at the national level.[74] There is a kind of political consensus, and the payments and credit-service providers, including India Stack and iSPIRT, are now considered part of the basic infrastructure of India, on which other infrastructures of health and education are seen to rely.

The methodological challenge in studying microloan apps is to connect the individual experiences of users at the frontend to the operation of fintech infrastructures in media backends. Given the volume of financial data gathered from transactions over the mobile internet in India today, describing and critiquing the FaaS infrastructure is a key part of studying mobile media distribution. It is a way to understand just how far the collection, verification, circulation, and processing of the phone-generated data of millions of Indians are extended from a single mobile phone. This FaaS infrastructure makes mobile internet finance possible, convenient, and also, on occasion, exploitative and anxiety-filled. Financial and social media data flows across geopolitical boundaries amid third-party players—including intermediaries like identity verifiers and risk scorers, who sniff, scan, process, nibble, and may even retain user data, unbeknownst not just to the user but also sometimes to the loan app and the nonbank lender. It is these flows of data across the mobile media distribution infrastructure that I have tried to conceptualize through the term "mobilities (un)limited" in several chapters of this book. What flows through these phones, distribution platforms, and circulatory infrastructures is not just data and content but also emotions and affect in new, mobile, mediated interactions.

Just the sheer number of stakeholders in the FaaS infrastructure also thoroughly contradicts fintech's lie (or promise) that that there are no intermediaries or human mediators between the borrower and the seller in digital lending. Intermediaries, both human and nonhuman, have increased in fintech operations. In the case of predatory loan apps, these intermediaries play crucial roles in shaping the shifting financial subjectivities of borrowers who suddenly realize that their user data have been compromised—that their parents, friends, and relatives know they are defaulters. The irony is that recovery agents might even know about the defaulter's inability to pay, and they are simply forcing defaulters to ask their relatives and friends for money. The recovery agents are themselves under pressure from their bosses, so they keep on with their mediated threats through calls and text messages. Along with data and money, what circulates in this fintech infrastructure are threats, negotiations, anxieties, and affect, leading to new and mediated unwanted intimacies, attachments, dependencies, and frictions.

Conclusion

Limiting: Surveillance, Environment, and Predatory Inclusions

Around 2016, buoyed by the so-called *data kranti* ("data revolution"), a significant customer base of "neomobiles," an aspirational neo–middle class of users/audiences in India (estimated to be more than two hundred million people and growing), acquired mobile phones and accessed the (mobile) internet for the first time in their lives. Neomobiles became the target for both the streaming media industry and fintech companies. Throughout this book, I have been asking the question, What kinds of digital infrastructures are emerging for content and money to move through and reach mobile phone users in contemporary India, and how does their design impact the forms of inclusions and exclusions enacted as well as the horizon of social behaviors and expectations in Digital India? This book, then, is about the connection between aspirational politics and mobile media distribution. Streaming/fintech services are as much concerned with infrastructural systems as they are with cultural form if we consider the multiple personhoods and experiences of neomobiles. "Neomobile" remains a contested term standing in for an idealized distribution subject of platformized governance or quasi-agential data generator in/of on-demand delivery cultures.

While the challenge for neomobiles is often made out to be "access" by the distributors of TV series and financial products, they themselves are facing the challenge of reaching existing audiences and discovering more audiences. Telecom, tech, and media firms are merging with and collapsing into one another in "search for audiences."[1] Cybersecurity expert Anand Venkatanarayanan writes that not just fintech start-ups and venture capitalists, but even scamsters have realized that "smartphones solve the distribution problem for a product" provided one has the crucial resource of phone numbers in order to reach and discover new customers or better still, new targets.[2] Scamsters have carried out their operations in crowded markets in the past and now have transitioned to digital platforms that allow broadcasting and promotion capabilities as well as discretion, pretense, concealment, and anonymity so as to monetize the targeted victim's vulnerability and launder (read: redistribute) money.

Several streaming platforms like Hotstar and SonyLiv worked out deals with telecom companies such as Reliance Jio and Airtel to have their services be part of bundled offers. Short-video platforms like Moj and Josh positioned themselves as shouldering post-TikTok futures in India by sending mobile stages on canter vehicles into the rural hinterlands for regional microinfluencers to showcase their talent. The state and fintech companies based on India Stack and FaaS infrastructures sought to include the until now unbanked Indian citizens into regimes of formal digital payments and lending through mobile phone apps. This provided the convenience of instantaneous transmission of monetary benefits to many customers as well as predatory possibilities for third-party loan app developers and mobile phishing operators to siphon off user data. Both platform studies scholars and distribution studies researchers acknowledge that there is a power tilt toward distribution in the media economy, as it intersects with the era of platforms and the dynamics of digital mobilities. This book offers an understanding of what the means, mode and forces of distribution really are.

Redrawing the Limits of Media Distribution

A key focus through the book has been on mobile media distribution, the ways in which on-demand delivery to mobile media devices is made possible through formal and informal intermediaries, infrastructural systems, and content curators. Another key concept threading across the chapters has been the theorization of mobilities (un)limited, the conjunctural moments where transportation modes (motorbikes, automobiles) come together in relation to the movement of content, data, and money into/through mobile media devices.[3] Mobilities (un)limited is very much about upward/downward socioeconomic mobilities as mobile phones become crucial sites to activate entrepreneurial content creation, to procure loans, and to send payments. This is not to suggest that mobile phones are being used exclusively for entrepreneurial and professional matters. In fact, as discussed in several chapters, given the cornucopia of multimedia objects and apps that have converged into the mobile phone, the device is used very much for entertainment and leisure pursuits, to pass time: people interact and get to know the world, and this interaction creates aspirations and drives mobilities. Even as mobile phones have become hyper-convergent devices with varied capabilities, they are not everything, and therefore, often I have moved out to explore their connections to wider distribution infrastructures, transportation systems, and the political economy that creates particular expectations about mobile phones, the affordances they provide, the hopes they can help fulfill.

In this book, I have not just pointed out what the limits of platformized mobile media distribution and aspirational politics are or why they exist or how they emerge, but also what the implications and ramifications of those limitations are for individuals, communities, and society. Loan apps in India, Kenya, and many others parts of the Global South arrive with the promise of both financial and social inclusion for marginalized poor people and profits for the apps' start-up ventures, but they do not address the contradictions and the "double binds" of some of the above-mentioned competing and seemingly incongruous logics.[4] Monitored behaviors (through digital traces of phone activity) are supposed to predict whether borrowers will be able to convert the monetary opportunities provided to them into success, but as Serena Natile has argued, such logics help expand the mobile money market but do not seriously engage with the "politics of redistribution," and thereby power, revenues and resources remain unequally distributed.[5]

It is not surprising that the video apps and multisided mobile money platforms that I discuss appear to break earlier physical and logistical constraints as they reach new users through their ability to scale and through network effects. While bureaucratic and corporate decision-making has involved demographic/address/property data before, the scale at and intensity with which platforms and computational algorithms are drawing on digital behavioral traces leads to extraction and creation of new values. Focusing on distribution backends helps specify how exactly platforms reintermediate. Loan disbursal is not one digital transaction, but rather, a set of connected transactions across platforms and network actors, many of these happening in the intermediate spaces of the multisituated fintech infrastructure that supports the loan app. The credit checks, authentication of identity, and loan approval that were once performed by human intermediaries are today being conducted by, or automated through, AI algorithms and SDKs. Indeed, one can say that the intermediation of lending platforms is a form of reintermediation in the lending process: Old intermediaries are being replaced by new intermediaries in money and data transaction chains.

The alternative data points on which the machine learning algorithms classify and judge new borrowers may not explicitly be based on gender or caste, but through their extreme behavioral focus, they do not necessarily eliminate gender/caste disparities; rather, they might be forking, splitting, dispersing, and branching such disparities further in myriad behaviors. The alternative datapoints deployed by the Fibe and CASHe loan apps may include how quickly users fill in online forms, how many mistakes they make, who their phone contacts are, what kind of premium lifestyle apps they have on their phone, where they shop, and how often they call their parents. Aggregating data across these touchpoints helps these apps put together what they call a "social loan quotient" (SLQ). The SLQ relates social behavior to long-time financial behavior.[6] Users are left guessing how to respond to (or game) this opaque process of determining creditworthiness (or intent to pay) in a digital financial landscape where there may not be stark binary digital divides (or financial exclusions) but what Jenna Burrell and Marion Fourcade phrase as a "sliding scale of classification situations" that "facilitate various forms of predatory inclusions."[7]

Aspirational Narratives: Speculations, Uncertainties, Histories

This book conceptualizes the role mobile media plays in the aspirational politics of contemporary India. Purnima Mankekar has explained that "aspirations are an affective-temporal formation: drawing sustenance from a disjuncture between how subjects navigate the present and imagine the future" and thus taking note of the continuities and disjunctures between the lived present and imagined futures.[8] Aspirations are not merely personal ambitions, for they lie in the contentious spaces between individual and collective desires.[9]

The mobile phone both constitutes and is constituted by aspirational mobilities. While this book acknowledges the power and possibilities of such aspirational mobilities, it critiques the "firmative speculation" of corporations and politicians that paints limited futures by stoking aspirations of the neo–middle classes. Such aspirational mobilities lack "affirmative speculation" that is open to radical uncertainties and differences.[10] Right-wing Hindutva nationalist politicians exhorting upper-caste Hindu men to express themselves on social media networks foreclose a future where radical religious differences can be acknowledged and sustained. Telecom corporations and platform companies offering apps for shopping for previously inaccessible consumer goods cut into the profits of and hence limit the future of small retail businesses in India as more consumers look to embrace the ideal neoliberal lifestyle. Narendra Modi's Digital India banks on futures of cashless transactions and mobile internet to such an extent that it eschews other "digital indias" of "low-level" underpaid software testing and call center work or the "unruly"/"improper" energies of thriving pirate cultures.[11] Digital start-ups and fintech enthusiasts promote the need for financial inclusion of "poor people" who live at the global margins, but keep mum when they are asked whose private corporate interests are being served by expanding Internet coverage and profiling poor households' data usage patterns.[12] These are the "limits" of mobilities (un)-limited, and they play a role in setting the terms of inclusion and exclusion.

The mobile phone is a part of the aspirational narrative in India, but it is equally true that there is a profound ambivalence about aspirational India. Some people I met had sold their agricultural fields and motorbikes in their village in the state of Bihar to come and make music in the city of Delhi (and to promote such music on their mobile phones). This involved considerable risk. Aspiration involves modeling growth trajectories by bracketing potential risks; aspiration sometimes even requires calculation, but not the methodological mapping of calculation. Therefore, the speculation involved in aspirational goals does not always try to tame uncertainty.[13] Any act of entrepreneurship, from opening a tea stall or a start-up, requires taking a bit of risk and engaging in speculation, and feelings of aspiration adjoin these acts. Aspirational subjects who take professional risks by leaving steady jobs to become dedicated content creators for a short-video platform, or financial risks in terms of taking out an online loan, are constructed as at once "risk-taking" subjects and "subjects-at-risk."[14]

Some microentrepreneurs note just how important the mobile phone (making quick calls or facilitating digital money transfers) is for their newly emerging business, from selling fish to driving cars (for ride-sharing). At the same time, there were others who used the phone (affording cheap data) to play games and for "timepass," because even though they had college degrees, they could not find (stable) work. Still other interview subjects

I met told me that since everybody aspires, what is the point of the aspirational optic I was espousing? When I asked Jamtara's locals about the stories of mobile phishing scamsters among the youths in their village, some told me their stories were illustrative of the aspirational narratives of small-town entrepreneurs, while others thought these youths were clearly criminals, although perpetrators of somewhat "soft" cybercrimes (posing as bank officials to divert customer monies to rented out Jan Dhan accounts). Academic friends I met in the United States told me that apart from class- and region-based differences, there was some similarity between Jamtara's scamsters and the deceptive CEOs of tech companies in Silicon Valley who duped their investors and walked away with millions. They seem to have inspired the "scam-trepreneur genre" and "tech TV dramas."[15] Journalist Samarth Bansal, who has reported widely on scams in India, stated rather matter-of-factly, "Whether we like it or not, scams are an outcome of entrepreneurship. We are at a time, if you think about the economy in general, it's a destabilized economy, widening inequality, and at these times, people are just craving for opportunities to make ends work."[16]

For some, like Deepu Sebastian Edmond, the journalist who played a pivotal role in the concept development and scripts of the Netflix show *Jamtara*, the dominance of aspirational discourses at a specific moment in the history of India has to be connected with the rise of neoliberalism in the country. This is when many thinkers and policymakers felt an urge to focus on class mobility without taking into account caste inequalities. According to Edmond, in the early 2010s, the aspirational framing became a way for upper-caste mainstream journalists to write about class and the Indian youth while avoiding caste considerations. The promise of the aspirational narrative in the heyday of neoliberal discourse held sway because that was a time when India was supposed to enjoy the demographic dividend of youth coming through the ranks, but that "demographic window has passed" Edmond notes, and with that, aspiration is not discussed with such enthusiasm as before.[17] Edmond's contention about people being less confident about brighter futures is shared by Harish Damodaran, who wrote in a recent column for *The Indian Express* that "aspiration is dead in India" and "the hope of scaling a higher economic and social position at a mass scale" is missing.[18]

According to Edmond, aspiration is a marketing idea that elides discussions of caste in considerations of class mobility. Edmond discerningly notes that caste perpetuates such entrenched hierarchies in Indian society that as a country, inequality in India is both starker than in other nations and has endured for a comparatively longer period of time. The reason for this has to be principally attributed to caste: "Primarily because of caste, we are locked in and aspiration is like a glass ceiling that is difficult to break through. And I think, a lot of people realize this even while we keep selling the idea of you know American dream and everything . . . but this idea of being locked in and we discuss that India is locked in a middle-income trap. It is a nice marketing trick to sell to people of a certain class and caste but in our heart of hearts we know that they are not going to escape this."[19] It is not surprising then that protagonists Sunny and Gudiya in *Jamtara*, despite being skilled with mobile phone technologies, find their aspirations to be repeatedly thwarted by an upper-caste politician.

Neoliberalism brought about the aspirational narrative, which proposed a way of thinking about caste-less class mobility in Inda, with some media outlets in early 2010s running programs containing titles such as *Caste, No Bar* or *Caste Set Aside by Aspiration.*[20] Ajeet Kumar Pankaj and Manish Jha have written about the "new Dalit middle class" in urban locations such as Navi Mumbai. This research indicates that Dalits feel more emancipated in urban locations and have found it easier to socialize, and yet the relevance and persistence of caste remains in the way they describe their struggles of getting a loan from a bank or renting an apartment when people find out their last names.[21] So, it is not that Dalits do not aspire for better education, for entrepreneurial success, and for class mobility; they definitely do. It is that even the new Dalit middle class in urban India still faces caste discrimination: They are discouraged from following entrepreneurial pursuits or attaining education goals by upper-caste populations. Anant Kamath found that historically deprived Dalit castes in peri-urban Bangalore could not take advantage of the opportunities provided by cell phone and apps to access new networks and contacts that could open the doors to jobs and enterprises. The already upwardly mobile Dalit castes were able to leverage the emerging mobile phone networks for business and livelihood opportunities and launch business ventures in Bangalore.[22] Here again, we find the importance of thinking about the entanglement of class and caste in understanding aspirational politics and mobilities (un)limited.

I have so far in this conclusion gestured toward some key takeaways that emerge while connecting several dots through the chapters. I will now engage with two key debates that continue to resonate in today's digital media scholarship and where I believe perspectives from the "Global South" are much needed. The first discussion topic concerns the access vis-à-vis surveillance debates that have arisen as more and more consumer-citizens are using mobile phones and are perceived to be submitting to governmental/corporate surveillance in exchange for ease of access. The second is about contestations regarding the ways the data that is increasingly made mobile (and undergoes wide circulation) is extracted, stored, and shared. Both these interrelated questions have implications for discourses about data autonomy, data sovereignty, data colonialism, and efforts to decolonize data when thought about in relation to nation-states, transnational corporations, consumer citizenship, and the distribution subject of the neomobile.

Access Vis-à-Vis Surveillance

Throughout the range of case studies that we have discussed in the book, various scenarios have emerged where there is a trade-off involved regarding consenting to surveillance (parting with user data) in exchange for access to services, whether it is opting-in for a loan app or downloading a messenger service. Over the last ten years, on my various fieldwork trips, I have heard government bureaucrats, tech journalists, and family friends say casually that caring about privacy concerns related to surveillance is a very "Western" idea. They note that people in India, especially the poor, and specifically those Indian neomobiles who are part of the world's next billion users on their way to becoming digital, are not bothered about privacy unless their own money gets stolen. Some, like tech journalist Shadma Shaikh, have reiterated to me that "the whole concept of privacy and surveillance is very foreign to most Indians" and added, "I do not think that any sort of privacy breach is taken seriously here. Ease of use of app is given more preference over security . . . nobody cares about whether the app is using my data as long as the app gives me good user interface. That is my impression, personal inference that I gather from the people that I speak with."[23] While Shaikh does provide a somewhat representative account, other researchers like Beni Chugh have suggested that India's lower-income groups are selective about what kinds of information they want to share and "care deeply about privacy of personal data"[24]

Payal Arora notes that poor internet users in the Global South, including India, might be accustomed to being watched by their neighbors, family, and the state, living as they do in one-room houses with their extended family, but that does not mean they are consenting subjects to digital surveillance all the time.[25] Arora explains that while some lower-income youths might not care too much about surveillance in their enthusiasm to interact on Facebook and WhatsApp, it is also possible that "this disenfranchised demographic has quickly learned the lay of the digital land" and "can invent strategies to circumvent surveillance."[26] The contextualization of privacy understandings is key here, as Arora mentions that youths in the small towns of India sometimes appreciate targeted advertisements, as these ads can be informative and save them time and money in buying the right product, which they otherwise have to keep looking for, but they draw the line when the ads are vulgar, as "they are concerned about self-respect."[27]

While users' mobile phones can be used to surveil them, there are various scenarios where users can deploy their mobile phone to surveil others. In 2015, the newly minted Aam Aadmi Party, which initially began as an anticorruption movement (India Against Corruption, led by Anna Hazare) against the corruption scams of the Congress-led United Progressive Alliance government and then became a political party (led by Arvind Kejriwal), went a step further. As Srirupa Roy writes, "Billboards and media advertisements" created by AAP "urged Delhi's citizens, to make audio or video recordings of incidents of bribe-taking by government officials and then turn in this 'evidence' and report the crime to an anti-graft helpline."[28] AAP even declared that it was making a "sting app" for citizens to secretly film corrupt dealings, where the phone screen would go dark, appearing to be switched off even as the phone camera was on and recording.[29] The "entrapment" videos associated with sting operations, which were mostly associated with either the police or media professionals, now became part of the activities of politicians, and even anticorruption citizen vigilante outfits such as auto-wallas entrapped the traffic police asking for bribes.[30] A certain mode and aesthetic of the sting video emerged not initially with the mobile phone but with Internet magazine/news-site Tehelka's March 2001exposé of a defense deal, Operation WestEnd, in which several members of the ruling National Democratic Alliance were caught on camera accepting bribes for favorable arms-related contracts.[31]

As mobile phones have become popular and ubiquitous, there also has been an increase in surveillance technologies, particularly those targeting mobile devices, ranging from the almost benign targeted advertising of some social media/video apps to lethal spyware like Pegasus. On July 18, 2021, Amnesty International and Forbidden Stories brought out a list of many politicians and journalists across the world who were (or had been) under the surveillance of the Israeli NSO group through the Pegasus spyware in their mobile phones. NSO mentioned that it has several nations' governments as clients, and it is those governments who use its Pegasus product to snoop on people, including intercepting all emails and texts (even encrypted ones). Pegasus spyware also makes it possible to on audio and video applications and silently track targets' locations.[32] The Pegasus issue very much dominated surveillance debates in India, as the list of names issued by Amnesty International included 121 politicians, journalists, and activists from India, a majority of them key opposition voices against the ruling dispensation (though a small minority were government ministers). The Indian government was alleged to have been a client of NSO.

Surveillance has become a crucial lens through which to understand the dynamics of citizenship in India with drives toward increasing digitization and biometricization and the entanglement of mobile media assemblages with these practices. In several chapters of the book, we looked at the Government of India's JAM initiative to connect Jan Dhan bank accounts with the Aadhaar card (unique identification card) and the mobile phone and several (un)intended consequences of that, including how mobile phishing scammers from rural India rented some of these new Jhan Dhan bank accounts from village elders (who themselves hardly used them) and deployed them to reroute money from the ATM cards of rich urban customers. Aadhaar is a key initiative to study for two reasons: (1) Its underlying premise is access to welfare benefits in exchange for "private" biometric information, and (2) the deployment of Aadhaar goes way beyond welfare delivery to the market function of being key to the fintech infrastructure that has been central to how money (and even content) moves through mobile phones. Aadhaar is therefore key to both the access vis-à-vis surveillance debates and to the circulation/mobility of data and money through mobile phones.

Aadhaar and India Stack at once can be considered digital distribution infrastructures as well as be understood to be digital public goods in that they call attention to the transparency and accountability required for the delivery of public goods and resources. Akhil Gupta writes that "infrastructures tell us about aspirations, anticipations, and imaginations of the future" which includes "what people think their society should be like, what they want it to be like, and what kind of statement they wish to make about that vision of the future."[33] When it was being built, Aadhaar was supposed to scale for a billion people given the population of India, and then with the imagination of stacking things, the plan was to make digital building blocks for payments, lending, and health on top of it. With Aadhar-enabled payment systems and India Stack, one sees a consensus among the Indian government policymakers, business, start-up tech entrepreneurs, and international financial institutions such as IMF and World Bank about financial inclusion premised on formal digital bank accounts, which then means that "providers of credit collectively have a basic infrastructural quality similar to providers of transport, electricity, or water supply systems."[34]

Aadhaar and India Stack as infrastructures and public goods help enact anticipatory governance for a particular vision of the future where seamless, efficient, and precise digital cash transfers (as indirect subsidies) can happen without the corruption and leakages that plagued earlier public distribution systems. It is this infrastructural vision that seemed to have succeeded when Pramod Varma, one of the chief architects of Aadhaar and India Stack noted that by November 2021, thanks to these digital infrastructures there were "650 million people" with unique bank accounts who were "financially addressable." According to Varma, this made it possible that "450 million people received subsidies directly in their bank accounts" while "America was writing checks and sending it to homes."[35] That the distribution of indirect subsidies or cash happened on digital rails in a developing country like India and took the form of mail and postal service in the developed nation of the United States indicates the national pride Varma felt in being more digitally/technologically sophisticated and in having a hand in building infrastructures that made the distribution of money instantaneous, efficient, and convenient. Of course, the terms of inclusion and exclusion that Aadhaar has enacted are very much debatable, and who is made accountable and who is left unaccountable in the distribution structures created by India Stack have sparked lots of controversies. The surveillance costs of the "unique addressability" of Aadhaar, so much touted by technocrats like Varma, also require further study.

Prior to Aadhaar, there were too many different documents that worked as identification papers and thus created fake identity issues. A need was therefore felt to assign every resident in India a unique identification number (UID/Aadhaar card). This uniqueness is based on an individual citizen's biometrics (fingerprints [all ten fingers], iris scans [of both eyes], picture of face), which could then be authenticated and verified online. Critics of Aadhaar note that it is based on biometrics, which cannot guarantee uniqueness, as duplicate entries remain.[36] They note that welfare delivery leakages happen not just at point-of-sale shops and hence that an efficient identification system cannot solve all problems of leakages.[37] In a celebrated dissenting opinion, Justice Chandrachud of the Indian Supreme Court contended that while Aadhaar by itself does not collect profiling information, it still carries the risk of profiling because Aadhaar can be potentially linked with other databases maintained by the state or private players.[38]

These debates about Aadhaar have been raging amid at least a partial reimagining of the state influenced by discourses of datafication and platformization. Emulating current trends of efficiency and optimization prevalent in private sector, the state seems to conceive itself as a platform.[39] Armed with Aadhaar, the state believes delivering government benefits/services to the mobile phone–equipped distribution subject (*e- labharthi/mSeva*) is the way to ameliorate poverty and expand the emerging neo–middle class. Suffice it to say, then, the beneficiary (*labarthi*) is imagined as comprising some portion of the neomobile category by the platform state. This is the new state-citizenship contract. While the state's stated goal of distributive justice is laudatory, one needs to be realistic about the extent of the improvement in delivery. According to a survey conducted to assess beneficiaries' access to government schemes, almost 40 percent of Jan Dhan account holders could not access the government's COVID-19 relief because of dormant accounts owing to infrequent use and limited access to banks.[40]

Highlighting such instances where intimate aspects of bodies have been converted into data, Anja Kovacs and Nayantara Ranganathan note that data can no longer be considered a resource; it is an extension of human bodies now.[41] Such questions about data, the body, and surveillance have been crucial in thinking about debates regarding data localization, which have often been framed in terms of the boundaries of the nation-state, but then what about boundaries around human bodies. These lead to profound questions about individual autonomy and national sovereignty in relation to data colonization.

While Aadhaar's role in the delivery of governmental subsidies and welfare benefits has been touted, Aadhaar has done much to create new mobilities of data in the many different market domains we have discussed in this book, including digital payments and lending. Making digital payments is crucial not just for the fintech sector but also for the streaming industry, with the subscriber video-on-demand business model in which subscribers need to pay in regular intervals. The biometric enacting of Aadhaar transformed the Indian resident's body into storable and retrievable data, and this instituting of "individuals as data points within databases" has not only served the so-called welfare beneficiary (*labarthi*) but also served the market players and data economy.[42] The identification, and then authentication infrastructures, that were created with Aadhaar, have then been made part of the India Stack APIs (as discussed previously), representing the consent layer and cashless layer. Through Aadhaar and the India Stack, the Indian state has been involved in deriving valuable data from the very large Indian population, and circulating that data to create more data, and therefore has been a participant in "enacting the logic of capital."[43]

Several scholars see Aadhaar as a data surveillance (dataveillance) exercise enacted by the Indian state in partnership with start-up tech entrepreneurs.[44] Some scholars have even gone further to suggest that the Indian government's data localization and data protectionist policies targeting foreign multinational platform players under the banner of digital sovereignty are a way for it to deflect attention from the data extraction practices involved in Aadhaar and India Stack.[45] The discussion of digital sovereignty is important because, so far, we have examined questions about the mobility of data within the country. That said, what regulatory limits can the Indian government place on the movement of data outside the nation's borders?

The massive amounts of Big Data gathered from mobile internet users find their way to data centers, some in India and some in other nations, and in the next section, I discuss the tenuous sustainability of such data-energy relations: Lots of energy is expended in maintaining data severs in these modern data warehouses. As I will illustrate, at the infrastructural level, storing Indian citizens' data in data centers in India may not always be equivalent to decolonization of data: Data extraction and energy extraction have to be considered more carefully together at both the micro- and macro-scales.

Locating Data Centers: Data-Energy Relations and Decolonizing Data

The storage location where content and data moving through mobile phones and various mobile media supply chain infrastructures are finally deposited is a data center. A data center could be considered the last, or first, node in the diagrammatic tracing of media backends of a mobile media distribution system. I will discuss questions of data surveillance, data sovereignty, and data–energy relations by elaborating on the physical, infrastructural siting of data centers in the city of Mumbai and its satellite city Navi Mumbai, a region that, since the 2000s, has witnessed major development of information technology parks.

"Global" Big Tech social media companies, fintech corporations, and video-on-demand streaming services are criticized for being data accumulators and aggregators and for implementing a regime of "data colonialism"/"digital colonialism" based primarily on data extraction as a new kind of capitalistic resource extraction. The data colonialism thesis described by Nick Couldry and Ulises Meijas states that while historical colonialism "appropriated land, resources, and bodies," today's "new colonialism" "appropriates human life through extracting value from data."[46] I complicate the aspects of "extraction" and "extractivism" embedded in the data colonialism thesis, where data at times simplistically end up being considered a resource.[47] Beyond just considering data as a resource, we need to understand the actual energy-related natural resources that are needed to store and circulate data.

Couldry and Meijas express concern that data have generally moved from the Global South to the Global North: the North-South data center traffic. Toward the end of the last decade (2010–2020), newspapers in India ran articles with titles such as "India Faces Threat of Digital Colonization."[48] Such articles compared the lobbying by contemporary social media companies and Big Tech giants to make sure that India does not have strict data localization laws (and that a free flow of data happened across India's borders) to the manipulative trade deals struck by the British East India Company in the eighteenth century, which in the long run led to India's colonization. This kind of journalistic reportage was happening alongside the introduction of an earlier iteration of India's Data Protection Bill in 2019, which had categorizations of personal data as "sensitive personal data," which could be sent outside the nation with governmental consent, and "critical personal data," which had to be processed in India.[49]

These provisions led to frenetic data center construction in Mumbai and Chennai, which are especially critical pressure points, as this is where the international fiber-optic landings happen.[50] However, the moist and warm weather in these places does not make it easy to keep the data centers cool, unlike the dry and cool air of the Scandinavian countries—a topic of discussion among scholars in both environmental media and media industry studies. That said, Mumbai and Chennai are (and have been) key industrial and financial hubs, which has meant that the local administration in these cities has experience in creating deregulatory environments in the form of relaxed land and labor laws, tax exemptions, and provisioning for scarce resources like water (and hydropower) to attract investments in the data center industry and other digital media industries.[51]

As digital capitalism expands in India, with mobile internet consumers adding data through e-commerce, gaming, streaming TV shows, and digital payments, the need for data centers is on the rise. Several international players, like Equinix and NTT, along with Indian companies like CtrlS, Yotta Infrastructure, and Sify Technologies, are investing significantly in data center construction and development. A report from the capital market company CRISIL started with the headline "Demand for Data Centers Goes Viral." With many people working from home, and hence Zooming for meetings, and watching more streamed entertainment, the pandemic seemed to have propelled the data center industry.[52] Citing India's massive data consumption amid COVID-19 and the growth of 5G technologies, a *Forbes India* story noted that data centers are a "real estate asset" that is "pandemic proof."[53]

Here is also an admission that data centers are key real estate. Data center owners, Daniel Greene writes, are "internet landlords," who collect rent for running specialized buildings equipped with state-of-the-art air-conditioning, power backup, and security systems, where racks and racks of data servers are located. Their tenants pay them for maintaining their digital assets in those servers, as well as for making their networks available for peering and interconnection.[54] Not surprisingly, many of the data center investors are either current or former telecommunications companies or real estate developers. The Hiranandani Group, one of India's leading builders, headquartered in Mumbai, is famous for having built bungalows, residential apartments, and shopping malls in various parts of the city, most famously in the Powai neighborhood. It should not come as a surprise that the Hiranandani Group transitioned in a big way to make hyper-scale data centers by backing a subsidiary, Yotta Infrastructure. Before venturing into the data center market by itself, Hiranandani had earlier built such infrastructures for the Japanese transnational company NTT or leased commercial spaces to them (see figure C.1).

Yotta's first big venture was in Navi Mumbai: the Yotta NM1 data center located in the six-hundred-acre Hiranandani Fortune City in Panvel (see figure C.2). While making my way there across a stretch of Navi Mumbai, I saw many billboards on the highway declaring that Hiranandani Fortune City had transformed Panvel into the next Powai (see figure C.3).

Powai is an upscale residential neighborhood in central Mumbai, while Panvel is about forty kilometers away. Panvel is witnessing development

Figure C.1
Area around NTT data center in Hiranandani building (Andheri, Mumbai). Photo taken by author.

as industries relocate to Navi Mumbai to escape the congestion in central Mumbai, yet the comparison struck me as strange, if not too boosterish. Some regions in Navi Mumbai, like Vashi and Belapur, much closer to Mumbai than Panvel, have experienced massive growth in the last three decades, initiated by the development of central business districts and the rise of information technology parks, including the Dhirubhai Ambani Knowledge City. While it has some petrochemical plants and textile-processing facilities, Panvel is more famous for rice fields, farmhouses, and lush green hills. Perhaps that is why it forms part of promised development projects including data centers, as it offers land not yet utilized for industrial and

Figure C.2
View of the Yotta data center from outside in Panvel. Photo taken by author.

Figure C.3
Billboards of Hiranandani real estate declaring Panvel is the next Powai. Photo taken by author.

real estate projects. Manish Jha and Rishi Jha note that Navi Mumbai has more potential for data center growth than Mumbai. Navi Mumbai's special economic zone is noncontiguous and fragmented, which allows data center projects to be geographically isolated from hazardous waste and provides green coverage that makes a contamination-free air supply possible for the data center buildings.[55]

Media infrastructures are associated with questions about energy transition and intersecting energy cultures. What one discovers from sifting through industry magazines and the business press covering the media and communication industry is that "green energy" is the way forward for data center companies. As energy guzzlers, data centers in India currently draw electricity from coal and fossil fuels. But that could be offset by data center operators either signing power purchase agreements to procure renewable energy or entering into partnerships with renewable energy developers to set up these projects. In the financial papers, sometimes one reads that a combination of solar and wind energy creates a stable grid, with solar functioning well in the day and wind taking over at night. Rarely are the photovoltaic cells key to harnessing solar energy discussed, and the energy-intensive process of producing them is almost never mentioned. Lithium, a key element in making renewable energy storage more affordable, is extracted from the salt flats of Argentina, Chile, and Bolivia, as Javiera Barandiarán has so meticulously discussed.[56] These lithium mines do not leave toxic tailing ponds like uranium mines do, but they require solar radiation to evaporate large amounts of water from the mineral-rich brine, taking away water from one of the driest ecosystems in the world.[57] Also, exactly where are the wind and solar farms located that are feeding electricity through the grid to these data centers of Mumbai and Chennai? What kind of land accumulation and infrastructural siting strategies have allowed for their operation? Several such examples exist where data centers are created in hyper-urban Mumbai, for which the energy comes from a solar plant or wind farm in semi-urban or rural places, like NTT simultaneously launching a data center in Mumbai and a solar plant in Solapur.[58]

As David Singh writes, the region of Kutch in the state of Gujarat is teeming with biodiversity, consisting of mangroves, grassland, thorn forests, and seasonal wetlands, but it has somehow been called a "wasteland" and appropriated for wind farm projects.[59] Wind farms are dispersed across many villages, grouped sometimes in clusters of about a dozen turbines. There is a cumulative process of land encroachment for transmission lines, pole stations, and networks of roads, making villagers feel that they are surrounded by windmills. Pamori, in Kutch, does not have a continuous electricity supply. While energy from Pamori served through the grid can hypothetically go to a data center in Ahmedabad or Mumbai, distributing streaming content (contained in those data centers) to Pamori might be difficult given that the edge of the node for CDNs like Netflix's Open Connect or Akamai ends in Mumbai or Ahmedabad, not beyond.

With vehement critiques of "data colonialism" in India, there is increasing demand for data centers to be housed within the country if they are storing data about Indian citizens. Data centers are energy guzzlers, and with rapidly depleting coal and oil reserves and concerns about climate change, data center companies are being asked to embrace postcarbon energy futures. This moment of energy transition to wind, solar, "nuclear," and geothermal energies has been contentious, and while there is genuine hope in renewables, there are also fears of "green capitalism" creating resource frontiers that not only are extractive but also devalue local knowledge systems and ways of living.

Finally, the talk of limitless mobilities has to meet the challenges of relatively sedentary (or immobile) infrastructures, like the data servers where most of the mobile internet data go. Media infrastructures like these work with energy sources that are scarce or difficult to procure, so there are limits to the mobile streaming story as well. What I hope to have made clear is that aspirational politics and the rhetoric around neomobiles frequently run into challenges of power and social divides, as mobile media distribution meets issues of data, energy, and capital. Data might seem like an infinite resource, but it does meet limits of energy and land. This dilemma or ambivalence about limitlessness and mobile media is what I have tried to convey through the concept of mobilities (un)limited throughout the book. While shifts in urban form, like the relocation and redistribution of data centers from Mumbai to Navi Mumbai do help in continuing the data center boom in India, challenges remain with regard to procuring green energy sources without engaging in problematic land accumulation practices and disrupting water ecosystems. The organization of mobile media distribution through data center networks encounters "the nexus between logistical media and environment" as well as the "blueprints of geopolitics."[60]

Neomobiles: Practices and Subjectivities

The story of access and surveillance can be told from the perspective of corporate land-and-resource grabs and from the perspective of biopolitical/geopolitical designs of the Indian state. It can also be told from the perspective of the everyday mobile digital practices of neomobiles in India, the challenges they face when they encounter app interfaces that have only recently begun to be redesigned to keep their contexts, their concerns, and their situation in mind. In the course of this book, I have tried to tell the story of mobile media in India from several interrelated perspectives: one about the different affordances that the new functionalities of cell phones offer; another regarding socio-technical practices of mobile media distribution systems that circulate content, money, and data through/across mobile phones; and still another perspective regarding how aspirational politics and mobile media systems interact with one another to influence who has what kind of access with what associated risks and challenges. Often, the information and communication technologies for development literature or mobile for development paradigm suggests that once there is access to information/mobile technologies, there will be empowerment, but having access and not having access is not a simple binary issue. It is a spectrum of differentiated accesses that one encounters in the field. Access to mobile technologies does not simply translate to agency or empowerment.[61]

Difficulties navigating text-dominated apps for emerging mobile users have to some extent been bypassed by voice inputs instead of text entry, but speech recognition algorithms have much catching up to do in terms of the wide variety of languages and diverse dialects spoken in India.[62] For first-time digital users, as designer Shrinath V. notes, logging in can seem like an intimidating gate, and the iconography of the floppy disk connoting save or the shopping cart for adding items could be alienating given that some of them have not seen a floppy disk or have not been to a shopping mall. The success of WhatsApp and TikTok among emergent mobile users in India had much to do with easy login and registration processes, and some apps have redesigned the "add items" icon in their apps to resemble a *theli* (conventional shopping bag).[63]

In the introduction, I discussed Sathana Chinnadurai's story of becoming a TikTok sensation from the remote Tamil village of Nachelura. Nilesh Christopher, who wrote one of the first journalistic pieces about Chinnadurai, mentioned to me in an interview that Chinnadurai's initial success was met with anxiety by her husband, who did not like hearing her dance moves being discussed by village men, including his friends, in tea stalls and *paan* shops. Throughout the history of cell phone adoption in India, gendered subjectivities and values assigned to particular mobile media practices have been a field of contention. Whether it is Sirpa Tenhunen writing about restrictions on newly married women calling their mothers from their in-laws' house in rural West Bengal or Anja Kovacs noting the constraints imposed by the *khap panchayat* in villages of Western Uttar Pradesh and Haryana regarding use of mobile phones by girls outside their houses, there is a recognition that erstwhile practices of the social surveillance of women have been challenged by women using mobile phones.[64] Some of these new restrictions or constraints imposed by in-laws or the village elders in the *khap panchayat* are ways to continue to retain control of women's movements, conversations, and interactions even as the use of mobile phones by women consistently defies such measures. Here again, we see recalibrations and regulations of mobilities afforded by mobile phone adoptions, something I have tried studying throughout this book with the concept of mobilities (un)limited.

There is sometimes an easy conflation of neomobiles and the neo–middle classes in India. The deployment of the term "neo-middle" by the BJP in several elections is both an exhortation to a key vote bank and a call for them to participate in the formalized market economy as a consumer/entrepreneur/producer.[65] Aspirations and desires tend to defy objective income-based categorizations, and they shape a subjective sense of belonging to a class group, often performatively expressed/enacted in intersubjective relations to members of same/other class groups.[66] These performative acts could be conspicuously watching Instagram Reels on a smartphone among friends in a public space or changing one's status or profile picture repeatedly in a WhatsApp group or driving a newly bought motorbike on a street. While Modi and his government might tout the promises of upward mobility to a secure middle-class position for neo-middle aspirers, it is important to point out that the neo–middle classes risk downward mobility and can fall into poverty, vulnerable as they are to economic downturns such as the pandemic, when social support systems broke down and there were massive job losses. Neomobiles therefore have a far more heterogenous class composition.[67]

Several other dimensions of caste, region, and gender intersect with class in shaping the fate of neo-middle aspirers and neomobiles as readers have found out in various chapters of this book. In 2015, I observed that security guards migrating from small towns and villages in Bihar, Jharkhand, and Orissa to cities like Bengaluru and Mumbai were living in tier 1 cities, but their listening practices are not similar to the wealthy youth smartphone-owners in such cities. When the security guard would go back home to Ranchi (in Jharkhand), he would get download vendors to download recent Bhojpuri or Sadri songs/music-dance videos on his phone to bring these songs with him to share with his group of security guard friends when he returned to Bengaluru. This significant migrant labor population defied any easy tier 1 and tier 3 categorizations. It is possible today that the same security guard has a modest smartphone, perhaps second-hand, and does not need memory cards to watch Bhojpuri music videos, as many of them are now available on YouTube, JioSaavn, and Moj. However, in 2015, as Rashmi M writes, it was important to understand that memory card downloading practices afforded migrant security guards, construction workers, and rickshaw drivers a key source of entertainment when the posh multiplexes in Bengaluru or Mumbai were simply outside their income range.[68] Today multiplexes themselves are not doing as good a business as they were before COVID-19, because the pandemic created a bigger audience base for SVOD (subscription video-on-demand)/AVoD (advertising-based video-on-demand) streaming apps. These neomobiles might be able to afford an economical package of apps bundled together for their entertainment on their second-hand Vivo Android smartphones. These same groups are being approached by loan management and social enterprises like Three Wheels United, which mediate between them and the banks and mobile money infrastructures like Novopay and Airtel Money.

This is a book about mobile phones in India, with a slant toward distribution. Studying mobile media distribution in relation to streaming and fintech is important, as these two industries are expanding the platform economy in India, designed as they are to push limits and enable the production of value whether through addictive viewing/attending eyeballs or credit scoring/debt traps so as to incorporate more and more neomobiles into the global capital market. The fintech ecosystem supporting digital payments and lending, with stakeholders including banks, telecom companies, and retailers, very much helped the subscriber video-on-demand services: The regular and repeated small-ticket transactions required for subscriptions to the video platform payment interfaces could now be supported in India without neomobiles requiring credit cards, which many of them simply did not have. Vinay Singhal, CEO of the regional OTT app Stage, which offers content in Haryanvi, Bhojpuri, and Rajasthani with significant customers in rural India, thanked "UPI's autopay" as a big "big life-saver" in their efforts to maintain customer retention and assess potential subscriber's intent. In Stage's efforts to become the "Netflix for Bharat" UPI's ability to support autopayment of such small amounts as "1 rupee" really helped.[69] The ability to contribute small monetary payments as fans of short video platform (Moj) influencers in the form of gifts to these micro-celebrities during their live-streaming app sessions also enabled a whole group of neomobile users to participate and feel included.

It is not only that everyday financial activities like payments and lending are being increasingly mediated through apps, but there is also an increasing convergence of payment platforms and media platforms.[70] A striking case of such convergence is livestreaming apps where streamers host a session interacting with their followers, and followers tap on virtual gifts to send to the host. Users/followers exchange rupees with the platform currency mints in the Indian livestreaming platform Moj Live and then use the mints to buy/gift flowers, planes, and Bollywood actor/character/song-dance icons like Baburao (from the film *Hera Pheri* [2020]) and Nattu Nattu (from the movie *RRR* [2022]), which appear in the form of pictographs and pop-up animations accompanied by music and live comments on the crowded mobile phone screen. The virtual gifts are convertible to cash, with the streamers and the platform splitting the revenue. A lot is going on here in terms of relational labor and interface design that scholarship on Chinese livestreaming platforms has discerningly pointed to. The mediation of money into livestreamed objects is key to tracing the circulation of capital in and across "livestreaming platforms [that] operate as payment platforms."[71] We find here the mutual constitution of digital content and money, where and when mediated content and money blur into and co-emerge with communicative interactions, activating phenomenal instantiations of mobilities (un)limited.

Mobilities (un)limited is about how various stakeholders such as media distributors and audiences, and digital lenders and borrowers make sense of the incongruities of mobile media distribution and more generally of life amid technocapitalism, and struggle, anticipate, speculate, live, and even thrive, through them. These above-mentioned examples might be the entertainment media and fintech industry's efforts to keep up with Indian youths' aspirational trajectories and find even more innovative/extractive monetization models, but many among the educated young people, who were supposed to be the "demographic dividend" of India, now exist merely as a precarious labor force, either unemployed or, at best, underemployed. Along with the risks of surveillance and environmental thresholds that cannot just be crossed, here we encounter further limits of the mobile phone story in India.

Notes

Introduction

1. "Hutch ka chota recharge -Irrfan Khan," Hutch ad, 2005, https://www.youtube.com/watch?v=909w0OK3FfY. Hutch was acquired by the telecom company Vodafone in May 2007, and later, Vodafone merged with Idea Cellular to form Vodafone Idea (Vi) in August 2018.

2. Jonathan Donner, "The Rules of Beeping: Exchanging Messages Via Intentional 'Missed Calls' on Mobile Phones," *Journal of Computer-Mediated Communication* 13, no. 1 (2008): 1–22.

3. Atul Bhattarai, "Don't Pick Up! The Rise and Fall of a Massive Industry Based on Missed Calls," *Rest of World*, April 5, 2021, https://restofworld.org/2021/the-rise-and-fall-of-missed-calls-in-india/.

4. I explain these techniques of bit-ladder optimization and deep-edge computing (also called "fog computing" or "mist computing") in detail in chapter 2. The term "logistics" was first deployed in 18th-century warfare, and then later, in 1960s, it became all-pervasive in neoliberal economics as the science of organizing space, time, and energy so as to move goods, people, and information as efficiently as possible to coordinate global supply chains and meet the demands of contemporary capitalism. See Matthew Hockenberry, Nicole Starosielski, and Susan Zieger (eds.), *Assembly Codes: The Logistics of Media* (Durham: Duke University Press, 2021), 2.

5. Jon Brodkin, "Verizon Wireless to Slow Down Users with Unlimited 4G LTE Plans," *Ars Technica*, July 25, 2024, https://arstechnica.com/information-technology/2014/07/verizon-wireless-to-slow-down-users-with-unlimited-4g-lte-plans/; also see Jatin Grover, "Reliance Jio and Bharti Airtel May Face TRAI Red Flag on Unlimited 5G," May 1, 2023, https://www.financialexpress.com/business/industry-reliance-jio-and-bharti-airtel-may-face-trai-red-flag-on-unlimited-5g-3069524/. On 4G plans, when subscribers exhaust their data limit before the new billing cycle, the Indian telecom service providers typically slow down data speeds to 64 Kbps.

6. Such choke points in mobile networked geographies are not only sites of congestion or blockage but also places where the power of regulatory, technical, and legal bodies is exercised. See Ronald Deibert, *Black Code: Inside the Battle for Cyberspace* (Toronto: McClelland & Stewart, 2013); also see Ashley Carse, Jason Cons, and Townsend Middleton, preface to "Chokepoints," special issue, *LIMN* 10 (2018): 2–3.

7. Chuck Tryon, *On-Demand Culture: Digital Delivery and the Future of Movies* (New Brunswick, NJ: Rutgers University Press, 2013).

8. Alexander Lewis Passah, "Internet Blackouts in Meghalaya: A Case of Emerging Complexities in the Digital Age," *Media, Culture & Society* 43, no. 8 (2021): 1515–1527; Bilal Hussain, "Kashmir Registers Highest Number of Internet Restrictions Globally," *VOA*, February 11, 2023, https://www.voanews.com/a/kashmir-registers-highest-number-of-internet-restrictions-globally-/6958516.html.

9. Ministry of Electronics and Information Technology, Government of India, Digital India, 2017. https://www.digitalindia.gov.in/

10. I certainly do not wish to make a presentist argument for in another time, Herbert Simon wrote, "The abundance of one resource causes a scarcity of another . . . Attention is the scarce factor in an information-rich society" and then went on to add, "There is little use in solving just one facet of the problem (energy or environment) without solving both." See Herbert Simon, "Managing in an Information-Rich World," in *Models of Bounded Rationality*, vol. 3: *Empirically Grounded Economic Reason*. (Cambridge, MA: MIT Press, 1997), 173–182. Furthermore, while marketers frequently use the ploy of "unlimited" to gesture to abundance, they also do resort to the tactic of declaring "limited-time"/"limited edition" offers to create a sense of scarcity and urgency so that customers have the incentive to rush to purchase.

11. Purnima Mankekar and Akhil Gupta, *The Future of Fururity: Affective Capitalism and Potentiality in a Global City* (Durham, NC: Duke University Press, 2025). It does bear noting that for the hypothetical questions asked at the Experimentarium exhibit, the opinion solicitation device was shaped in the form of a mobile phone. I used scare quotes around "We" because I believe there is a need to find imaginaries and visions of futures at a scale more nuanced than species.

12. Pablo Boczkowski, *Abundance: On the Experience of Living in a World of Information Plenty*, (Oxford: Oxford University Press, 2021), 15.

13. Timothy Cooper, *Moral Atmospheres: Islam and Media in a Pakistani Marketplace* (New York: Columbia University Press, 2024), 113.

14. Simran Jain, interview by the author, July 12, 2024. While Simran certainly does not use "cringe" that way, it is important to note that the "cringe" label has been pejoratively deployed for creators from sociohistorically underprivileged backgrounds, those who struggle to meet the aesthetic standards of privileged and bigoted creators and gatekeepers on platforms like Instagram and YouTube.

15. Simran Jain, interview by the author.

16. Lauren Berlant, *Cruel Optimism* (Durham, NC: Duke University Press, 2011).

17. Cooper, *Moral Atmospheres*. Such a spatio-temporal politics of limits/thresholds is understood by both individuals and corporations. Capitalistic ventures also hold on to limits and even declare them when it pertains to minimizing risks in the future. For example, "limited warranty" periods on sold products considering their wear and tear, even planned obsolescence, and "limited liability" for "Private Limited" companies.

18. Orthodox upper-caste Brahmans opposed the entry of Dalits and women to schools, fearing this would threaten their monopoly over education and administrative jobs in pre-Independence India. Historian Shailaja Paik documents that even when Dalit girls were allowed admission to schools, they had to sit at the door/threshold of the classroom. See Shailaja Paik, *Dalit Women's Education in Modern India: Double Discrimination* (New York: Routledge, 2014), 48.

19. Kathryn Zyskowski, "Aspiration as Labour: Muslim Women at a Basic Computer-Training Centre in Hyderabad," *South Asia: Journal of South Asian Studies*, 43, no. 4 (2020): 758–774.

20. Vincanne Adams, Michelle Murphy, and Adele E. Clarke. "Anticipation: Technoscience, Life, Affect, Temporality," *Subjectivity* 28, no. 1 (2009): 246–265.

21. Karishma Desai, "Immaterial Precarity and Affective States of Anticipation: NGO Afterlives of Non-Elite Young Women in Urban India," *Anthropology and Education Quarterly* 54, no. 1 (2023): 6–21.

22. https://education.economictimes.indiatimes.com/amp/news/how-women-in-indias-hinterlands-are-driving-the-skilling-revolution/118734013

23. Desai, "Immaterial Precarity." There has been some criticism about personality development camps and life skills training programs organized by both the Indian government and social justice activists as being part of the neoliberal discourse of individual self-making. That said, Savitha Suresh Babu has noted that the general wholesale critique of such life skills programs as merely instilling a neoliberal idea of confidence culture might be too hasty. Confidence conceptualized as self-respect by anticaste movements can interact with confidence understood as a soft skill considered to be key for succeeding in private-sector jobs and enterprises. The self-making personality development program that Suresh Babu attended in Bangalore encouraged Bahujan students to view their lives in relation to caste structures that are as much a part of the socio-religious worlds as they are of the market economy in India. So, it is possible that remaking of selves is not individualized but connected to the political collective of Bahujan students. See Savitha Suresh Babu, "Education for Confidence: Political Education for Recasting Bahujan Student Selves," *South Asia: Journal of South Asian Studies* 43, no. 4 (2020): 741–757.

24. Megan Finn and Mark Ananny, "Making Events: How Anticipatory Infrastructures Produce Shared Temporalities," *New Media & Society* 3 (2024). OnlineFirst.

25. Alisa Perren, "Rethinking Distribution for the Future of Media Industry Studies," *Cinema Journal* 52, no. 3 (2013): 167.

26. Emily West, *Buy Now: How Amazon Branded Convenience and Normalized Monopoly* (Cambridge, MA: MIT Press, 2022); Joshua Braun, "News Distribution," in *Oxford Research Encyclopedia of Communication*, ed. J. F. Nussbaum (New York: Oxford University Press, 2021).

27. Courtney Brannon Donoghue, Timothy Havens, and Paul McDonald, "Introduction: Media Distribution Today," in *Digital Media Distribution: Portals, Platforms, Pipelines*, ed. Paul McDonald, Courtney Brannon Donoghue, and Timothy Havens (New York: NYU Press, 2021).

28. Platformization of distribution services did not begin with Apple and Android phones. Marc Steinberg argues that an earlier model of platformization was developed through iMode/DOCOMO phones in late 1990s Japan, setting up innovative relations between content and money, mobile internet, and platforms. See Marc Steinberg, *The Platform Economy: How Japan Transformed Commercial Internet* (Minneapolis: University of Minnesota, 2019). I first encountered the phrase "global mobile media" in a book by Gerard Goggin who suggested looking at "cell phones as a species of global media" and pointed out how cellular phones could help us understand and reframe how different communication and media technologies relate to one another. See Gerrard Goggin, *Global Mobile Media* (London and New York: Routledge, 2011, 3).

29. "2022 Internet in India," Kantar, https://www.iamai.in/sites/default/files/research/Internet%20in%20India%202022_Print%20version.pdf.

30. It must be noted that for the purposes of this book, I am using "mobile money" mostly for the movement of money across connected mobile digital devices such as mobile wallets and payment and lending apps. That said, an expanded definition and set of examples for mobile money should include "airtime," where prepaid mobile airtime minutes can be exchanged between phones. Unlike the payment app kind of mobile money where the telecom network is being used as a channel for cash, "airtime" is a new currency that "directly transacts on the value of these ethereal (telecom spectrum) channels." See Rachel O'Dwyer, "Ether," in *Paid: Dongles, Checks, and other Money Stuff*, ed. Bill Maurer, Lana Swartz, and Bruce Sterling (Cambridge, MA: MIT Press, 2017), 239. In the early 2010s, prepaid airtime minutes could be swapped for cash or spent in retail stores in Egypt, Ghana, Uganda, and Nigeria.

31. Adrian Athique and Vibodh Parthasarathi, "Platform Economy and Platformization," in *Platform Capitalism in India*, ed. Adrian Athique and Vibodh Parthasarathi (London: Palgrave MacMillan), 12.

32. Lana Swartz, *New Money: How Payment Became Social Media* (New Haven, CT: Yale University Press, 2020). Also see Shayan Momin, "Lumpen Utopia: Money,

Livestreaming, and Labor in Emerging Media Worlds," *Journal of Cultural Economy* 16, no. 2 (2023): 221–241.

33. Nimmi Rangaswamy, "Disruptive IT in South India," in *Sarai Reader 3: Shaping Technologies*, ed. Ravi Vasudevan, Ravi Sundaram, Jeebesh Bagchi, Monica Narula and Shuddhabrata Sengupta, no. 3 (2003): 170–176.

34. Rich Ling and Heather Horst, "Mobile Communication in the Global South," *New Media & Society* 13, no. 3 (2011): 363–374.

35. See the association of mobile network operators GSMA's *Unlocking Rural Coverage: Enablers for Commercially Sustainable Mobile Network Expansion. Connected Society* (London: GSMA, 2016).

36. Daniel Greene, *The Promise of Access: Technology, Inequality, and the Political Economy of Hope* (Cambridge, MA: MIT Press, 2021).

37. Assa Doron and Robin Jeffrey, *The Great Indian Phone Book: How the Cheap Cell Phone Changes Business, Politics, and Daily Life* (Cambridge, MA: Harvard University Press, 2013); also see Jo Tacchi, "Being Meaningfully Mobile: Mobile Phones and Development," in *Technological Determinism and Social Change: Communication in a Tech-Mad World*, ed. Jan Servaes (London: Lexington Books, 2019), 110.

38. Savita Bailur, Silvia Masiero, and Jo Tacchi, "Gender, Mobile, and Development: The Theory and Practice of Empowerment. Introduction," *Information Technologies & International Development* (special section), 14 (2018): 96–104. Jeffrie et al. note that while in India as of 2022, there is 11% difference in phone ownership between men and women, there is 40% difference in mobile internet use. Also, 19% of female internet users are accessing it on somebody else's phone compared to 8% of men. This suggests women may be using more of basic phones vis-a-vis smartphone. See Nadia Jeffrie et al., "The Mobile Gender Gap Report 2023," Global System for Mobile Communications Association, https://www.gsma.com/r/wp-content/uploads/2023/07/The-Mobile-Gender-Gap-Report-2023.pdf

39. Jo Tacchi and Tripta Chandola, "Complicating Connectivity: Women's Negotiations with Smartphones in an Indian Slum," in *Routledge Handbook of New Media in Asia*, ed. Larissa Hjorth and Olivia Khoo (London: Routledge, 2019), 179–188.

40. Jo Tacchi, Kathi Kittner, and Kate Crawford, "Meaningful Mobility: Gender, Development and Mobile Phones," *Feminist Media Studies* 12, no. 4 (2012): 528–537.

41. Lending company professional working in Mumbai, interview by the author, June 4, 2024.

42. Lisa Parks, Julia Velkova, Dander de Ridder, eds. introduction to *Media Backends: Digital Infrastructures and Socio-Technical Relations* (eds.) Lisa Parks, Julia Velkova, and Dander de Ridder (Urbana-Champaign: University of Illinois Press, 2024), 1–22.

43. Ravi Sundaram, "Hindu Nationalism's Crisis Machine," *HAU: Journal of Ethnographic Theory* 10, no. 3 (2020): 734–741.

44. Bishnupriya Ghosh, "Big Bad Social Media: Distributed Affects and Popular Politics," *Media Populism: Special Issue of Cultural Politics* 19 (2020), https://culturemachine.net/wp-content/uploads/2021/04/1.-Bishnupriya-Ghosh-revised-1.pdf.

45. Sahana Udupa, Shriram Venkatraman, and Aasim Khan, "Millennial India: Global Digital Politics in Context," *Television & New Media* 21, no. 4 (2020): 343–359.

46. Tarunima Prabhakar, "Downloading a Debt Trap," *Indian Express*, January 18, 2021, https://indianexpress.com/article/opinion/digital-lending-loan-apps-rbi-banks-7151613/.

47. Swartz, *New Money*

48. These are discourses one can find in global management consulting group McKinsey's collection of essays, *Reimagining India: Unlocking the Potential of Asia's Next Superpower.*"

49. Vijayanka Nair, "Governing India in Cybertime: Biometric IDs, Start-Ups, and the Temporalized State," *South Asia: Journal of South Asian Studies* 42, no. 3 (2019): 519–536.

50. Darin Barney, ""We Shall Not Be Moved": On the Politics of Immobility," In *Theories of the Mobile Internet: Materialties and Imaginaries*, ed. Andrew Herman, Jan Hadlaw, and Thom Swiss (New York: Routledge, 2014), 19.

51. Doron and Jeffrey, *The Great Indian Phone Book.*

52. The undermining of social hierarchies that Doron and Jeffrey discuss has been criticized by scholars like Jo Taachi and Tripta Chandola, who find the authors to be overchampioning mobile technologies' disruptive potentials and overcrediting its power in social transformation at the expense of local appropriation of mobiles. See Tacchi and Chandola, "Complicating Connectivity."

53. Noah Arceneaux and Anandam Kavoori, "Introduction: Mapping Mobile Media," in *The Mobile Media Reader*, ed. Noah Arceneaux and Anandam Kavoori (New York: Peter Lang, 2012), 2.

54. Cara Wallis, *Technomobility in China: Young Migrant Women and Mobile Phones* (New York: NYU Press, 2013).

55. Wallis, *Technomobility in China.*

56. Mara Mills, "The Audiovisual Telephone: A Brief History," in *Handheld? Music Video Aesthetics for Portable Devices*, ed. H. Keazor (ART-Dok 2012), 34–47; also see Jason Farman, "Historicizing Mobile Media: Locating the Transformations of Embodied Space," in *Mobile Media Reader*, ed. Noah Arceneaux and Anandam Kavoori (New York: Peter Lang, 2012). Portability and mobility differ with regard to ease of movement and connectivity.

57. These *cable wallas* operate in the hyperlocal neighborhood and have witnessed several socio-technical transformations in the past: from video parlors to cable TV,

and from cable TV to digitalization/encryption of broadcasting signals requiring setup boxes.

58. Chuck Tryon, *On-Demand Culture*; Swartz, *New Money*.

59. Lana Swartz and David Stearns, "Money and Its Technologies: Making Money Move in the Modern Era," in *A Cultural History of Money in the Modern Age*, ed. Taylor C. Nelms and David Pedersen (London: Bloomsbury Academic, 2019), 27–52.

60. Sibel Kusimba, *Reimagining Money: Kenya in the Digital Finance Revolution* (Stanford, CA: Stanford University Press, 2021).

61. Jennifer Pybus and Mark Cote, "Did You Give Permission? Datafication in the Mobile Ecosystem," *Information, Communication & Society* 25, no. 11 (2022): 1650–1668.

62. Rahul Mukherjee, "Jio Sparks Disruption 2.0: Infrastructural Imaginaries and Platform Ecosystems in 'Digital India,'" *Media, Culture & Society* 41, no. 2 (2019): 175–195.

63. In this Scroll.in article, former finance minister Arun Jaitley commented on neo–middle class and aspirational India. See "Budget Gives Financial Direction to Aspirational India, Says Arun Jaitley," *Scroll*, July 6, 2019, https://scroll.in/latest/929643/budget-gives-political-direction-for-aspirational-india-says-arun-jaitley.

64. Tereza Capelos and Ipshita Basu, "Who Is in the Middle: Social Class, Core Values, and Identities in India," *Political Psychology* 43, no. 1 (2022): 89. Modi thanked this neo–middle class group when he won the Gujarat state elections in 2012, and he then vigorously used the term during the 2014 national elections. See Ullekh NP, "Targeting Neo Middle Class Was Narendra Modi's Key Game Plan," *Economic Times Bureau*, December 21, 2012, https://economictimes.indiatimes.com/news/politics-and-nation/targeting-neo-middle-class-was-narendra-modis-key-game-plan/articleshow/17699885.cms?from=mdr.

65. Arjun Appadurai, "The Capacity to Aspire: Culture and the Terms of Recognition," in *Culture and Public Action: A Cross-Disciplinary Dialogue on Development Policy*, ed. Vijayendra Rao and Michael Walton (Palo Alto, CA: Stanford University Press, 2004), 59–84.

66. Ellen Barry, "Hopes of a Generation Ride on Indian Vote," *New York Times*, May 5, 2014, https://www.nytimes.com/interactive/2014/05/15/world/asia/india-election.html. Beyond slotting people into middle class based on income, consumption patterns, and behavioral measures, Kapur et al. note the importance of self-identification. 47% of lower middle-income respondents in their study self-identified as middle-class. They found middle-class self-identification to be a strong predictor of "social aspiration" and "economic optimism." See Devesh Kapur, Milan Vaishnav, and Nilanjan Sircar, "The Importance of Being Middle Class in India," in *The New Middle Class in*

India and Brazil: Green Perspectives?, eds David Damilo Bartlet and Axel Harneit-Sievers (Academic Foundation and Heinrich Boll Foundation, 2018), 39–58.

67. Appadurai, "The Capacity to Aspire," 67.

68. Nimmi Rangaswamy and Payal Arora, "The Mobile Internet in the Wild and Every Day: Digital Leisure in the Slums of Urban India," *International Journal of Cultural Studies* 19, no. 6 (2016): 616.

69. Rangaswamy and Arora, "The Mobile Internet," 620.

70. Mimi Sheller and John Urry, "The New Mobilities Paradigm," *Environment & Planning A* 38 no. 2 (2006): 207–226; Leah Koskimaki and Carol Upadhya, "Introduction: Reconsidering the Region in India: Mobilities, Actors and Development Politics," *Journal of South Asian Development* 12, no. 2 (2017): 89–111.

71. David Nieborg and Thomas Poell, "The Platformization of Cultural Production: Theorizing the Contingent Cultural Commodity," *New Media & Society* 20, no. 11 (2018): 4275–4292.

72. Nilesh Christopher and Varsha Bansal, "Meet the Rising Stars of India's Video Apps," *Rest of the World*, May 12, 2020, https://restofworld.org/2020/india-video-apps-influencers/.

73. Brooke Erin Duffy, "The Romance of Work: Gender and Aspirational Labour in the Digital Culture Industries," *International Journal of Cultural Studies* 19, no. 4 (2016): 441–457.

74. Julie Soleil Archambault, *Mobile Secrets: Youth, Intimacy, and the Politics of Pretense in Mozambique* (Chicago: University of Chicago Press, 2017), 22.

75. Nandini Gooptu, ed. introduction to *Enterprise Culture in Neoliberal India: Studies in Youth, Class, Work and Media* (New York: Routledge, 2013), 8.

76. Priya Chacko, "Gender and Authoritarian Populism: Empowerment, Protection, and the Politics of Resentful Aspiration in India," *Critical Asian Studies* 52, no. 2 (2020): 204–225. The BJP government's social policies are aimed toward dilution of labor protection regimes and reduction of government jobs and socioeconomic rights-based inclusion in favor of initiatives that champion self-employment through entrepreneurial projects, promote ease of doing business, and encourage market dependency through financialization (financial inclusion through easy loan) projects.

77. Press release for Niti Aayog's Vocal for Local initiative: https://www.pib.gov.in/PressReleasePage.aspx?PRID=2014411.

78. Several Marxist scholars have noted that Karl Marx initially had a moralizing association of the "lumpenproletariat" with crime, homelessness, and degeneracy, which he later replaced with the more scientific "relative surplus population," which connoted a population that for various reasons was separated from capitalist

production, and hence was forced to find criminal and parasitical means of living. For more on this, refer to J. Sakai's work, which uses "lumpen/proletariat" to suggest that the lumpenproletariat and the proletariat are not that distinct but rather are related to each other. J. Sakai, *The "Dangerous Class" and Revolutionary Theory: Thoughts on the Making of the Lumpen/Proletariat* (Montreal: Kersplebedeb, 2017).

79. Anurag Dwary, "Unemployed? No, They Are "Aspirational Youth," Says Madhya Pradesh," *NDTV*, March 27, 2025, https://www.ndtv.com/india-news/unemployed-no-they-are-aspirational-youth-says-madhya-pradesh-802560. This is an example of an irresponsible political statement that made some Indian citizens cynical of aspirational talk, and increasingly a kind of aspiration fatigue seems to be setting in. Politicians as this book shows do set aspirational goals that are lofty as a way to motivate and mobilize people, affirm identities and shared values. Political scientists have noted that politicians do not face censure or condemnation if they fail to realize aspirational targets provided they can demonstrate having made significant efforts toward reaching those goals. See Martha Finnemore and Michel Jurkovich, "The politics of aspirations," *International Studies Quarterly*, 64 (2020): 759–769.

80. Peter Van der Veer, "The Future of Utopia," *History and Anthropology* 27, no. 3 (2016): 254.

81. For a detailed analysis of the views of this particular content creator, Kajal Chauhan, see Rahul Mukherjee, "Aspirational Politics of Talent Acquisition: Entrepreneurial Limits and Indian Short Video Platforms," *Social Media + Society*, April–June (2025): 1–12.

82. Donoghue, "Introduction," 6. As much as these individuals are conditioned by technologies and corporate power, they also exert their own agencies in order to ensure optimized delivery: Conversing with them about distribution processes that from afar often appear to be technical and structured begins to seem more contingent. These contingencies, whether they are decisions regarding rerouting of information packets or shifting the location of cache servers from one cell antenna to another, are then incorporated and automated over time as part of distribution logistics.

83. Braun, "News Distribution."

84. Kavita Dattani, "'Governtrepreneurism' for Good Governance: The Case of Aadhaar and the India Stack," *Area* 52, no. 2 (2019): 411–419.

85. Jacqueline Hicks, "Digital ID Capitalism: How Emerging Economies Are Reinventing Digital Capitalism," *Contemporary Politics* 26, no. 3 (2020): 330–350.

86. Braun, "News Distribution."

87. Perren, "Rethinking Distribution"; Lobato, *Shadow Economies of Cinema.*

88. Ramon Lobato, "The Cultural Logic of Digital Intermediaries: YouTube Multichannel Networks," *Convergence* 22, no. 4 (2016): 348–360; Smith Mehta, "Indian

Intermediaries in New Media: Mainstreaming the Margins," *Convergence* 27, no. 1 (2021): 229–246.

89. Ishita Ghosh and Jacki O'Neill, "The Unbearable Modernity of Mobile Money," *Computer Supported Coop Work*, 2020, https://link.springer.com/article/10.1007/s10606-020-09373-1. They also note that many such agents contributed to the success of a mobile money scheme (M-PESA) in Kenya, though celebratory declarations often champion the mobile handset as the one and only reason behind the transformation to a cashless economy.

90. André Brock, *Distributed Blackness: African American Cybercultures* (New York: NYU Press, 2020), 139.

91. Archambault, *Mobile Secrets*, 21.

92. Diana Zulli and David James Zulli, "Extending the Internet Meme: Conceptualizing Technological Mimesis and Imitation Publics on the TikTok Platform," *New Media and Society* 24, no. 8 (2022): 1872–1890.

93. Ravi Sundaram, "Introduction: The Horizon of Media Studies," in *No Limits: Media Studies from India*, ed. Ravi Sundaram (Delhi: Oxford University Press, 2013), 3.

94. As the 2019 McKinsey Global Institute report attests, "competitive offerings by telecommunications firms have turbocharged internet subscriptions and data consumption, which quadrupled in both 2017 and 2018 and helped bridge a digital divide; India's lower-income states are growing faster than higher-income ones in internet infrastructure and subscriptions."

95. Rashmi M., "Offline Media Economy: Digitally Marginalized Users of Mobile Phones," in *The Indian Media Economy*, vol. II, *Market Dynamics and Social Transactions*, ed. Adrian Athique and Vibodh Parthasarathi (Oxford: Oxford University Press, 2018), 208–209.

96. Abhishek Baxi, "The Rise and Rise of China's Xiaomi in India," *Forbes*, September 12, 2017, https://www.forbes.com/sites/baxiabhishek/2017/09/12/the-rise-and-rise-of-chinas-xiaomi-in-india/.

97. Sajith Pai, "India 1, Avocado Startups, and Product-Market Fit," *Medium*, February 18, 2019, https://medium.com/hackernoon/india1-avocado-startups-product-market-fit-dbfb7a8b2ef6. Some apps like YouTube and WhatsApp do cut across various user segments/tiers.

98. Raghavendra Prajapati, interview by the author, December 19, 2022.

99. Shrabonti Bagchi, "Sharechat: The Indian Language Social Network That Gets Bharat in a Way Facebook Doesn't," *FactorDaily*, October 24, 2016, https://archive.factordaily.com/indian-language-social-network-gets-bharat-way-facebook-cannot/.

100. Nilesh Christopher, interview by the author, February 12, 2021.

101. Pradip Thomas, *The Politics of Digital India: Between Local Compulsions and Transnational Pressures* (Oxford: Oxford University Press, 2019), 10.

102. Mukherjee, "Jio Sparks Disruption 2.0."

103. Rohin Dharmakumar, "Platform Ambitions: The Story of How iSpirt Lost Its True North," The Ken, September 21, 2017.

104. Aria Thaker, "The New Oil: Aadhaar's Mixing of Public Risk and Private Profit," *Caravan Magazine*, April 18, 2018.

105. Adrian Athique, "A Great Leap of Faith: The Cashless Agenda in Digital India," *New Media & Society* 21, no. 8 (2019): 1697–1713.

106. Antara Das, "Interview: Nandan Nilekani on 'Rebooting India,'" *Swarajya*, November 23, 2015.

107. While Tarleton Gillespie has focused on the "content moderation" roles of platforms, José Van Dijck has explained the architecture and expansion of platforms through strategies of "datafication" (capturing users' data and metadata), "commodification" (of digital goods into transactional value), and "selection" (of user's activity). Tarleton Gillespie, *Custodians of the Internet: Platforms, Content Moderation, and the Hidden Decisions That Shape Social Media* (New Haven, CT: Yale University Press, 2018); José Van Dijck, Thomas Poell, and Martijn de Wall, *The Platform Society: Public Values in a Connective World* (London: Oxford University Press, 2018).

108. Akshaya Kumar, "Informality in the Age of Platformization," *Media Industries* 6, no. 2 (2019), https://quod.lib.umich.edu/m/mij/15031809.0006.207?view=text;rgn=main.

109. Adrian Athique, Vibodh Parthasarathi, and SV Srinivas, *Indian Media Economy: Industrial Dynamics and Cultural Adaptation* (Delhi: Oxford University Press, 2019).

110. William Mazzarella, "Beautiful Balloon: The Digital Divide and the Charisma of New Media in India," *American Ethnologist* 37, no. 4 (2010): 785.

111. For more, refer to Nilesh Christopher, "As WhatsApp Pay Enters India, Local Fintech Companies Aren't Happy," *Rest of World*, November 13, 2020, https://restofworld.org/2020/whatsapp-pay-enters-india/.

112. In his speech inaugurating the PMJDY bank account scheme, Narendra Modi infused the neoliberal technocratic view of financial inclusion with Hindu religious and social values. He equated the entry of unbanked poor into the formal "economic system" (*arthvyavastha*) to removing "financial untouchability." Here, he invoked Mahatma Gandhi's fight against caste-based "social untouchability" and noted that PMJDY was a fight against poverty. Narendra Modi, "PM Modi's Speech at the Launch of the 'Pradhan Mantri Jan Dhan Yojana,'" YouTube, August 28, 2014, https://www.youtube.com/watch?v=7wUmxR10GO8; Ismail Erturk Indradeep Ghosh, and

Kadambari Shah, "Financial untouchability: A Polysemic Narrative of Digital Financial Inclusion in Modi's India," *Journal of Cultural Economy* 15, no. 1 (2022): 30–51.

113. Ananya Roy, "Subjects of Risk: Technologies of Gender in the Making of Millennial Modernity," *Public Culture* 24, no. 1 (2012): 131–155; Silvia Federeci, "From Commoning to Debt: Financialization, Microcredit, and the Changing Architecture of Capital Accumulation," *South Atlantic Quarterly* 113, no. 2 (2014): 231–244.

114. The crises of microfinance busts have not stopped the microfinance industry from finding new ways of making debtors out of the poor in the name of "financial inclusion" through "digital inclusion" in the form of fintech companies and microloan apps. Gabor and Brooks term this the "fintech-philanthropy-development complex." See Daniela Gabor and Sally Brooks, "The Digital Revolution in Financial Inclusion: International Development in the Fintech Era," *New Political Economy* 22, no. 4 (2017): 423–436.

115. Beni Chugh, workshop discussion at Dvara Research, Mumbai, August 18, 2024.

116. There are limits to efficiency and optimization, as Rossiter and Zehle put it evocatively: "Logistical media" by design "strive to organize environments," but "contingency lurks within the fissures of environment. Entropy is always on the heels of control." Ned Rossiter and Soenke Zehle, "Logistical Media and Environmental Media: Energy, War, and Politics of Distribution," *Journal of Environmental Media* 4, no. 1 (2023): 78. Ezra Klein and Andrew Thompson propose solutions to convert the scarcity of housing and energy into a state and politics of abundance of such resources and building blocks even noting that more energy is required to run data centers so that United States has the lead in the world AI race. AI searches and the actions of machine learning algorithms are indeed energy intensive. Their solutions though seem techno-utopian narrowing the discussion to state-market relations where the state mostly deregulates or provides subsidies, without considering that so-called green capitalisms are also marked by contradictions and extractivisms. See Ezra Klein and Andrew Thompson, *Abundance*, NYC: Simon & Schuster, *2025*.

Chapter 1

1. The Bhojpuri-speaking region (BSR), spread across western Bihar, eastern Uttar Pradesh, and northwestern Jharkhand, is again not specifically regional. Bhojpuri does not have one region or state, for there are Bhojpuri migrants in Mumbai, in Delhi, and in rural agricultural sites in Punjab. Kathryn Hardy notes that contemporary Bhojpuri filmmakers shift Bhojpuri cartographies to address Bhojpuri migrants across the country. See Kathryn Hardy, "Constituting a Diffuse Region: Cartographies of Mass-Mediated Bhojpuri Belonging," *BioScope* 6, no. 2 (2015): 145–164. This chapter draws significantly from two research trips I conducted with Abhigyan Singh in Mewat and Gaya. We have together published articles that are referred to here: Rahul Mukherjee and Abhigyan Singh, "MicroSD-ing 'Mewati

Videos': Circulation and Regulation of a Subaltern-Popular Media Culture," in *Asian Video Cultures: In the Penumbral of the Global*, ed. Joshua Neves and Bhaskar Sarkar (Durham, NC: Duke University Press, 2017), 133–157; Rahul Mukherjee and Abhigyan Singh, "Reconfiguring Mobile Media Assemblages: Download Cultures and Translocal Flows of Affective Platforms," *Asiascape: Digital Asia* 4 (2017): 257–284. While I do borrow from these earlier cowritten articles, I have added new conversations in this chapter and recent scholarship on memory card and USB drives (flash memory cards) from both India and Cuba to make connections about download cultures and alternative mobile media distribution systems.

2. According to Ratnakar Tripathy, Bhojpuri cinema, songs, and music videos can be observed as "self-assertion and empowerment of Bhojpuri language" that is "straining to rise above its 'dialect' status into a fully-fledged language with substantial cultural lineage and wealth." See Ratnakar Tripathy, "Mapping the Invisible World of Bhojpuri Cinema and Its Changing Audience," in *Routledge Handbook of Indian Cinemas*, ed. K. Moti Gokulsing and Wimal Disanayake (New Delhi: Routledge, 2013), 151. Bhojpuri music is lyrical—that is, it is lyric-intensive and language matters. It is sometimes religious, sometimes political, often about love and displacement (particularly in its *Bidesia* refrain), sometimes erotic (*Sringar Ras*), and known for its *Quawalli* structure and groovy percussion beats (*Theka*).

3. I refer to memory cards as "microSD cards," "mSD cards," and "chips" interchangeably in this chapter.

4. Musicologist Peter Manuel has updated his work on vernacular music practices in India by discussing the shift from cassette culture to VCD culture. I would argue that since 2009 the memory card culture also needs to be seen as interwoven with the VCD culture. See Peter Manuel, "Popular Music as Popular Expression in North India and the Bhojpuri Region, from Cassette Culture to VCD Culture," *South Asian Popular Culture* 10, no. 3 (2012): 223–236.

5. Swati Janu, "On-the-Go Settlements: Understanding Urban Informality Through Its Digital Substructure." *Urbanisation* 2, no. 2 (2017): 116–134.

6. Rashmi M., "Offline Media Economy: Digitally Marginalized Users of Mobile Phones," in *The Indian Media Economy*, vol. II, *Market Dynamics and Social Transactions*, ed. Adrian Athique and Vibodh Parthasarathi (Oxford: Oxford University Press, 2018), 202–219.

7. Ramon Lobato, "Postcolonial Piracy: Media Distribution and Cultural Production in the Global South," in *Postcolonial Piracy*, ed. Lars Eckstein and Anja Schwarz (London: Bloosmbury Academic, 2014). China and India have often been considered the world capitals of piracy, and such piratical citadels are then framed as impeding the growth of viable global cultural industries, such as (predictably) Hollywood. A formal recorded music industry has also existed in India with copyright regulations, where singers, lyricists, composers, and performers receive royalties. In some

cases, the formal and the informal get tangled together, and there are musicians who move across the formal and informal domains. For more on this, see Aditya Lal, David Hesmondhalgh, and Charles Umney, "The Changing Shape of the Indian Recorded Music Industry in the Age of Platformisation," *Contemporary South Asia*, 2023, https://doi.org/10.1080/09584935.2023.2206998.

8. It is important to note the difference between mSD and SIM (subscriber identity module) cards. An mSD card is a removable flash memory card used for storing data in mobile phones and other electronic devices (like digital cameras and audio players). A SIM card is used for storing a mobile subscriber's identity and telephony.

9. Neha Kumar and Tapan Parikh, "Mobiles, Music, and Materiality," paper presented at the *SIGCHI Conference on Human Factors in Computing Systems* (CHI'13), Paris, France, 2013, accessed April 5, 2017, http://dl.acm.org/citation.cfm?id=2481396/.

10. Rashmi M., "The Inchoate Field of Digital Offline: A Reflection on Studying Mobile Media Practices of Digital Subalterns in India," *Digital Culture and Society* 3, no. 2 (2017): 219–227.

11. Laura-Zoe Humphreys, "Utopia in a Package? Digital Media Piracy and the Politics of Entertainment in Cuba," *Boundary 2* 49, no. 1 (2022): 231–262; Steffen Kohn, "Unpacking El Paquete: The Poetics and Politics of Cuba's Offline Data-Sharing Network," in *Digital Culture & Society*, ed. Inequalities and Divides in Digital Cultures, 5 no.1 (2019): S. 105–124.

12. Ratnakar Tripathy categorically refuses to equate the "vernacular" with the "traditional," arguing instead that the term "vernacular" stands for a "shifting combination of the traditional and the non-traditional forms and practices employed by the music industries." See Ratnakar Tripathy, "Music Mania in Small-Town Bihar: Emergence of Vernacular Identities," *Economic and Political Weekly* 47, no. 22 (2012): 58–66. Also see Akshaya Kumar, "The Aesthetics of Pirate Modernities: Bhojpuri Cinema and the Underclasses," in *Arts and Aesthetics in a Globalizing World*, ed. Raminder Kaur and Parul Dave-Mukherji (London: Bloomsbury, 2015), 185–203.

13. Aditi Deo and Vebhuti Duggal, "Radios, Ringtones, and Memory Cards or, How the Mobile Phone Became Our Favourite Music Playback Device," *South Asian Popular Culture* 15, no. 1 (2017): 41–56.

14. Ravi Sundaram, "Post-Postcolonial Sensory Infrastructure," *e-flux* 64 (2015), http://www.e-flux.com/journal/64/60858/post-postcolonial-sensory-infrastructure/.

15. Jayson Beaster-Jones, "Music Piracy, Commodities, and Value: Digital Media in the Indian Marketplace," in *The Oxford Handbook of Mobile Music Studies*, vol. I, ed. Sumanth Gopinath and Jason Stanyek (Oxford: Oxford University Press, 2014), 434–455.

16. Beaster-Jones, "Music Piracy."

17. Tarleton Gillespie, "The Politics of 'Platforms,'" *New Media & Society* 12, no. 3 (2010): 347–364.

18. Jose Van Dijck, "Public Values in a Platform Society," keynote presentation at the Association of Internet Researchers Conference, 2016. Accessed April 5, 2017, https://www.youtube.com/watch?v=-ypiiSQTNqo/.

19. Annabelle Gawer, "Bridging Differing Perspectives on Technological Platforms: Toward an Integrative Framework," *Research Policy* 43, no. 7 (2014): 1239–1249.

20. Bruno Latour, *Reassembling the Social: An Introduction to Actor-Network Theory* (New York: Oxford University Press, 2005), 39.

21. Latour, *Reassembling the Social*, 118.

22. Karen Barad, *Meeting the Universe Halfway: Quantum Physics and the Entanglement of Matter and Meaning* (Durham, NC: Duke University Press, 2007).

23. Neha Kumar, "The Gender Technology Divide or Perceptions of Non-Use," *First Monday* 20 no. 11 (2015), http://journals.uic.edu/ojs/index.php/fm/article /view/6300/5133/.

24. By asserting this, I do not discount media practices in the United States, where students copy movies from the external hard disks of other students or people share music or films using USB sticks or Airdrop. The distinction is that in the United States such a downloading culture has not taken the form of local and informal business economies, as in the case of Mewat and Bihar in India, which I discuss in this chapter.

25. Quoted in Samuel Hansell, "Blockbuster's Big Gulp of Technology," *New York Times*, December 8, 2008, http://bits.blogs.nytimes.com/2008/12/08/blockbusters -big-gulp-of-technology/?_r=0/. Netflix has historically used the "unlimited"label to characterize its services to emphasize the large amount of content it makes available: it has suggested "Unlimited Streaming: Instantly" in the past or streaming "Unlimited movies, TV shows, and more" recently.

26. Hansell, "Blockbuster's Big Gulp."

27. Robert Wilonsky, "'Do You Like to Do It Yourself?' How Blockbuster Ended Up in the Self-Service Biz," *Dallas Observer*, November 10, 2009, http://www .dallasobserver.com/news/do-you-like-to-do-it-yourself-how -blockbuster-ended-up-in-the-self-service-biz-7141553/.

28. Neha Kumar, "Mobiles, Media, and the Agency of Indian Youth," PhD diss., UC Berkeley, 2013.

29. James Pearson, "The $50 Device That Symbolizes a Shift in North Korea," *Reuters*, accessed March 27, 2015, https://www.reuters.com/article/technology/the -50-device-that-symbolizes-a-shift-in-north-korea-idUSKBN0MM2UY/.

30. Pearson, "The $50 Device."

31. Deo and Duggal, "Radios, Ringtones, and Memory Cards," 2017.

32. Kumar, "Mobiles, Media, and the Agency."

33. Shail Mayaram, *Against History, Against State: Counterperspectives from the Margin* (New York: Columbia University Press, 2003).

34. There is a severe scarcity of water in Mewat—many times, when interviewees hosted us for lunch, conversations emerged about how they had to deal with an uneven water supply on an everyday basis. The large portion of the rural population in Mewat lives in utter poverty and is plagued by a lack of basic infrastructure. Many Meo men in Salaheri complained that development funds allocated for the economically backward region of Mewat were siphoned off by corrupt officials or used to further develop the already developed districts of Haryana.

35. Swati Chattopadhyay and Bhaskar Sarkar, "Introduction: The Subaltern and the Popular," *Postcolonial Studies* 8, no. 4 (2005): 357–363.

36. Mukherjee and Singh, *MicroSD-ing Mewati Videos.*

37. Chattopadhyay and Sarkar, "The Subaltern and the Popular," 359

38. Chattopadhyay and Sarkar, "The Subaltern and the Popular," 360.

39. Gayatri Chakravorty Spivak, "Scattered Speculations on the Subaltern and the Popular," *Postcolonial Studies* 8, no. 4 (2005): 475–486. The study of the intersection of subaltern and popular is attentive to the political landscape changing with new medial affordances. And yet, here, I am not trying to conflate "agency" with "empowerment."

40. Since my research in 2012, much has changed in the gig economy, and Mewat and other parts of peri-urban and rural Haryana close to New Delhi and Gurugram have been brought into the fold of the transportation business. There are drivers from Mewat who have taken loans and bought cars and drive them for Uber and Ola in Delhi. See Gayas Eapen, "Mobile Media Practices in Suburban North India," PhD diss., North Carolina State University, 2023. Eapen's dissertation examined not only the differential mobilities in the gig economy of Haryana but also mobilities involved with street travelers (*Yatras, Khel*) and DJ trucks. Meos have hustled in the face of severe income inequality and economic disparity in Nuh and participated in phishing operations, which I will discuss in chapter 4. The association of Mewatis with crimes has continued in the register of cybercrime.

41. On transregional mobilities, see Leah Koskimaki and Carol Upadhya, "Introduction: Reconsidering the Region in India: Mobilities, Actors and Development Politics," *Journal of South Asian Development*, 12, no. 2 (2017): 89–111.

42. K. C. Chamberlain, "Makin' It in School: How At-risk Kids Stay in High School and Maintain Optimistic Aspirations," *Explorations* 4 (2001): 1–42; Saskia Witteborn,

"The Digital Gift and Aspirational Mobility," *International Journal of Cultural Studies* 22, no. 6 (2019): 1–16.

43. Witteborn, "The Digital Gift and Aspirational Mobility," 4.

44. Mukherjee and Singh, "Reconfiguring Mobile Media Assemblages."

45. Tripathy, "Mapping the Invisible World."

46. The mobilities literature have attended to how the movement and stasis of bodies are inextricably tied with media practices and how they impact and are impacted by identities and subjectivities, something I have tried to borrow from and extend in this book. For more, see Adriana De Souza e Silva and Jordan Frith, *Mobile Interfaces in Public Spaces: Locational Privacy, Control and Urban Sociability* (New York: Routledge, 2012).

47. Tim Cresswell, "Towards a Politics of Mobility," *Environment and Planning D: Society and Space* 28, no. 1 (2010): 26.

48. The anthropologist Lotte Hoek once mentioned to me that she encountered Hindu devotional songs being circulated using memory cards in Bangladesh. Media anthropologist Brian Larkin noted that people who download Nollywood videos are called "downloaders" in Nigeria (akin to download vendors in India).

49. Such a distribution assemblage sparked by memory sticks transformed urban sociality, shaped entrepreneurial cultures, and led to the influx of K-pop in Cuba. Refer to Humphreys, "Utopia in a Package?" 235.

50. Humphreys, "Utopia in a Package?" 242. The *paquete* succeeded as a pirate, circumvention, and distribution technology not just for technical reasons but also due to the discourses about the social role of art and media.

51. Kohn, "Unpacking El Paquete; Humphreys, "Utopia in a Package?"

52. Geoffrey Hobbis, The MicroSDs of Solomon Islands: An Offline Remittance Economy of Digital Multi-Media, *Digital Culture & Society* 3, no. 2 (2017): 21–40.

53. Debra Spitulnik, "Alternative Small Media and Communicative Spaces," in *Media and Democracy in Africa*, ed. Goran Hyden Michael Leslie, & Folu F. Ogundimu (New Brunswick, NJ: Transaction, 2002), 177–206.

54. The new micro-SD-enabled Coke/Pepsi-bottle audio players represent contemporary digital media in places in India where streaming internet content was not possible until at least 2016. These are digital platforms and economies that operate in places that are "extranet"—that is, outside the internet. See Joshua Neves and Bhaskar Sarkar, introduction to *Asian Video Cultures: In the Penumbral of the Global*, ed. Joshua Neves and Bhaskar Sarkar (Durham, NC: Duke University Press, 2017). For instance, these microSD audio players can be found in mobile phone shops that also sell solar-panel-enabled mobile phone chargers because such shops are located in regions that are off the electricity grid in India. This context is critical

for understanding the dynamism of emerging digital platforms in the "developing world," where people continue to express themselves in a world not of a digital divide but of differentiated digital access.

55. Jackie O'Neill, Kentaro Toyama, Jay Chen, Berthel Tate, & Aysha Siddique, "The Increasing Sophistication of Mobile Media Sharing in Lower-Middle-Class Bangalore," paper presented at the *Eighth International Conference on Information and Communication Technologies and Development* (ICT4D 2016), Ann Arbor, MI, accessed June 13, 2025, https://dl.acm.org/doi/10.1145/2909609.2909656.

56. Shadma Shaikh, "From Sharing to Streaming: SHAREit's run to stay at the top in India," *FactorDaily*, January 8, 2019, https://archive.factordaily.com/from-sharing-to-streaming-shareits-run-to-stay-at-the-top-in-india/.

Chapter 2

1. Lea Zouien, "OTT in Africa: Can Mobile Operators Transform Their 550M Internet Users into Streamers?" *Dataxis*, September 12, 2023, https://dataxis.com/researches-highlights/1564774/ott-in-africa-can-mobile-operators-transform-their-550m-internet-users-into-streamers/.

2. Philippe Bouquillion, Christine Ithurbide, and Tristan Mattelart, introduction to *Digital Platforms and the Global South: Reconfiguring Power Relations in the Cultural Industries*, ed. Philippe Bouquillion, Christine Ithurbide, and Tristan Mattelart (London: Routledge, 2024), 1–17.

3. A more creator-friendly environment emerged on the Internet beginning with the success of The Viral Fever (TVF) Media Lab, which started an online YouTube channel. Their web series *Permanent Roommates* debuted in 2014 and went on to become a huge success. This encouraged the flight of creative talent to Internet content creation: Those who earlier either could not break into mainstream Bollywood films or found the television landscape to be pedestrian decided to collaborate to make web series. See Smith Mehta, "Indian Intermediaries in New Media: Mainstreaming the Margins," *Convergence* 27, no. 1 (2021): 229–246.

4. The merger of Sony and the Zee group was discussed starting in August 2023, which would have also led to a merger of the media companies' streaming services, SonyLIV and Zee5. However, the merger talks were called off in January 2024.

5. Shanti Kumar and Aswin Punathambekar, "Hotstar: Reimagining Television Audiences in Digital India," in *From Networks to Netflix: A Guide to Changing Channels*, ed. Derek Johnson (London: Routledge, 2022).

6. Reliance Jio was soft-launched on December 27, 2015, and then publicly made available on September 5, 2016. With earlier investments in textiles and petrochemicals, Reliance Industries now decided to focus on the telecom sector. Adrian Athique

and Akshaya Kumar, "Platform Ecosystems, Market Hierarchies and the Megacorp: The Case of Reliance Jio," *Media, Culture & Society* 44, no. 8 (2002): 1420–1436.

7. Earvin Charles B. Cabalquinto, *(Im)mobile Homes: Family Life at a Distance in the Age of Mobile Media* (London: Oxford University Press, 2022).

8. Samina Mishra, "Dish Is Life: Cable Operators and the Neighborhood," in *Image Journeys: Audio-Visual Media and Cultural Change in India*, ed. Christiane Brosius and Melissa Butcher (New Delhi: Sage, 1999), 261–278.

9. Dal Yong Jin, "The Construction of Platform Imperialism in the Globalization Era," *Communication, Capitalism & Critique* 11, no. 1 (2013): 145–172. Also see Scott Fitzgerald, "Over-the-Top Video Services in India: Media Imperialism after Globalization," *Media Industries* 6, no. 2 (2019).

10. Adrian Athique, *Transnational Audiences* (Cambridge: Polity Press, 2016).

11. Philippe Bouquillion and Christine Ithurbide, "Audio-Visual Industry and Digital Platforms in India: A Contribution from Political Economy of Communication," *Global Media and Communication*, 2022, 1–19.

12. That said, for its computing needs, Netflix does use Amazon Web Services (AWS) and therefore does have structural dependency. However, Netflix Open Connect Appliances help Netflix to maintain significant control of the distribution supply chain, something I will explain in greater detail later in this chapter.

13. Joshua Braun, "Points of Origin: Asking Questions in Distribution Research," in *Digital Media Distribution: Portals, Platforms, Pipelines*, ed. Courtney Brannon Donoghue Courtney, Timothy Havens, and Paul McDonald (New York: NYU Press, 2021), 42.

14. Braun, "Points of Origin."

15. Lisa Parks et al., eds., *Media Backends: Digital Infrastructures and Socio-Technical Relations* (Urbana-Champaign: University of Illinois Press, 2023).

16. Ramon Lobato, *Netflix Nations: The Geography of Digital Distribution* (New York: NYU Press, 2019), 100–101.

17. Jonathan T. Caldwell, *Production Culture: Industrial Reflexivity and Critical Practice in Film and Television* (Durham, NC: Duke University Press, 2008). Also see Courtney Brannon Donoghue et al., "Introduction: Media Distribution Today," in *Digital Media Distribution: Portals, Platforms, Pipelines*, ed. Courtney Brannon Donoghue et al. (New York: NYU Press, 2021).

18. Rahul Vaz, interview by Rahul Mukherjee, August 6, 2022.

19. Devika Narayan, "Monopolization and Competition Under Platform Capitalism: Analyzing Transformations in the Computing Industry," *New Media & Society* 25, no. 2 (2023): 287–306. Also see Akshay Sudhir and Prashanth Rao,

"How Technology is Shaping the Streaming Services in India," *Financial Express*, January 11, 2021, https://www.financialexpress.com/business/brandwagon-how-technology-is-shaping-the-future-of-streaming-services-in-india-2162729/.

20. Rahul Vaz, interview.

21. Rahul Vaz, interview.

22. Shanti Kumar and Aswin Punathambekar, "Hotstar: Reimagining Television Audiences in Digital India," in *From Networks to Netflix: A Guide to Changing Channels*, ed. Derek Johnson (London: Routledge, 2022). Also see "India Watch Report," Hotstar, 2019, https://secure-media.hotstar.com/media-website/indiawatchreport2019/India_Watch_Report_2019.pdf.

23. Ethan Tussey, *The Procrastination Economy: The Big Business of Downtime*, (New York: NYU Press, 2018).

24. Samar Khan, interview by Rahul Mukherjee, August 12, 2022, Mumbai.

25. Raymond Williams, *Resources of Hope: Culture, Democracy, Socialism* (London: Verso, 1989).

26. When this interview was conducted with Ekta Kapoor, she was the CEO of ALTBalaji, but in February 2023, she stepped down from her leadership role. Ekta Kapoor, interview with *Money Control*, July 28, 2020, https://www.youtube.com/watch?v=Vw4KlgUxoqM.

27. Ekta Kapoor made her foray into the Indian television scene in the mid-1990s, producing TV shows revolving around generational tensions in traditional joint family settings. Such soap operas included *Kyunki Saas Bhi Kabhi Bahu Thi* (Because Mother-in-Law was once Daughter-in-Law). A significant change occurred when Kapoor opened the OTT banner with ALTBalaji (alternative to Balaji), which moved away from traditional family melodramas of Balaji telefilms toward edgy action and erotic content around the sexual lives of small town and village characters epitomized by its flagship show, *Gaandi Baat/Dirty Talk: Urban Stories from Rural India*. While tier 2 and tier 3 towns remain the key audience demographic for ALTBalaji, it would be hypocritical to assume that the upper-middle-class audience does not watch *Gaandi Baat*, as research studies found people watching the show using high-end digital devices as well. "ALTBalaji's 'Gandii Baat' Enjoys Popularity Across Demographics," *Indian Television*, January 17, 2020, https://www.indiantelevision.com/iworld/over-the-top-services/altbalaji-s-gandii-baat-enjoys-popularity-across-demographics-200117.

28. Danish Khan, interview by Rahul Mukherjee, September 2, 2022, Mumbai.

29. Danish Khan, interview.

30. Sohini Mitter, "How Telco Partnerships Spurred the Growth of Netflix, Amazon Prime Video, ALTBalaji and Others in India," *YourStory*, February 11, 2019, https://yourstory.com/2019/02/netflix-amazon-altbalaji-hotstar-airtel-partnerships.

31. Bouquillion and Ithurbide, "Audio-Visual Industry"; Vibodh Parthasarathi, Philippe Bouquillion, and Christine Ithurbide, "Cross-sectoral Relations in VoD Markets: Frontend, Backend, and Deepend in India," In Lisa Parks et al. *Media Backends: Digital Infrastructures and Sociotechnical Relations*, Chicago: University of Illinois Press; Athique and Kumar, "Platform Ecosystems."

32. Using an aggregator model for its content services, Jio has added local and regional TV channels such as Mahua Plus (for Bhojpuri speakers), Assam TV (for consumers in the state of Assam who speak Assamese), and Manorama News (for people in the state of Kerala who speak Malayalam). See Imon, "30 New Channels Added on JioPlay App," *TelecomTalk*, August 29, 2016, https://telecomtalk.info/30new-channels-added-on-jio-play-app/156536/#!parentId.

33. Mitter, "How Telco Partnerships."

34. This kind of terminology does give one pause as to the telecom's oligopolistic implications. Super apps are indeed do-everything apps that bring an expansive suite of services within a single interface. Examples of such super apps include WeChat in China and LINE in Japan, which incorporate multiple functions such as chatting, texting, and mobile payments within a single app. The convenience of one price and one login are apparent. However, in thinking about convenience, what also needs to be mentioned is how super apps are realized only through powerful mega corporations: WeChat in China cannot be imagined without its parent company Tencent, and LINE has the backing of SoftBank—both of which are very powerful oligopolistic players. See Marc Steinberg, Rahul Mukherjee and Aswin Punathambekar, "Media Power in Digital Asia: Superapps and Megacorps," *Media, Culture & Society* 44, no. 8 (2022): 1405–1419.

35. Adarsh Nair quoted in TechDesk, "Airtel Xstream Premium 'Streaming Super App' with 15 OTT Services Announced," *Indian Express*, February 10, 2022, https://indianexpress.com/article/technology/airtel-xstream-premium-streaming-super-app-with-15-ott-services-announced-7765459/.

36. Nair quoted in TechDesk, "Airtel Xtream Premium."

37. Hotstar CEO Ajit Mohan quoted in Mitter, "How Telco Partnerships."

38. "India Watch Report 2019"; "Indian Media and Entertainment Landscape," FICCI Frames 2018

39. TechDesk Indian Express, "Airtel Xtream Premium"; LiveMint, August 28, 2022.

40. BS Web Team, "Reliance Jio Removes Disney+ Hotstar Bundle Offer from Select Prepaid Plans," *Business Standard*, October 14, 2022, https://www.business-standard.com/article/technology/reliance-jio-removes-disney-hotstar-bundle-offer-from-select-prepaid-plans-122101400278_1.html.

41. Kumar and Punathambekar, "Hotstar: Reimagining Television Audiences."

42. On Akamai's site about its customer Hotstar, see "Akamai and Hotstar Create History During the Final of Vivo IPL 2018 by Delivering 10.3 Million Concurrent Streams," https://www.akamai.com/resources/customer-story/hotstar.

43. Jean K. Chalaby and Steve Plunkett, "Standing on the Shoulders of Tech Giants: Media Delivery, Streaming Television and the Rise of Global Suppliers," *New Media & Society* 23, no. 11 (2021): 3206–3228.

44. Shankar, the former Asia-Pacific president of Disney and the former chairman of Star and Disney India, quit Disney in late 2020. Shankar is widely credited for Hotstar's streaming success and for inking the historic Rs 16,347.5-*crores* bid for the TV and digital rights of IPL in 2017 while at the helm of Star India. After leaving Disney (and Star India) in 2020, Shankar worked on his investment firm Bodhi Trees, which has since bought a stake in Viacom18 (whose majority shares are owned by Reliance's Network18), and Shankar himself is most likely to join Viacom18 not just as an investor but as a key leader and operator. Harveen Ahluwalia, "Understanding Reliance, Viacom18, and Uday Shankar," *Morning Context*, April 29, 2022, https://themorningcontext.com/business/understanding-reliance-viacom18-and-uday-shankar.

45. Harveen Ahluwalia and T. Surendar, "Disney's IPL Loss is Reliance's Gain," *Morning Context*, June 15, 2022, https://themorningcontext.com/business/disneys-ipl-loss-is-reliances-gain.

46. Bouquillion and Ithurbide, "Audio-Visual Industry."

47. See Manish Singh, "Disney's Hotstar Loses 8.4 Million Subscribers as Ambani's JioCinema Ascends," *TechCrunch*, May 11, 2023.

48. Abhimanyu Mathur, "JioCinema and Disney+ Hotstar Merge to Form New Platform JioHotstar, Announce New Subscription Plans," Hindustan Times, February 14, 2025, https://www.hindustantimes.com/entertainment/web-series/jiocinema-and-disney-hotstar-merge-to-form-new-platform-jiohotstar-announce-new-subscription-plans-101739502610427.html.

49. Lobato, *Netflix Nations*.

50. Chalaby and Plunkett, "Standing on the Shoulders."

51. Chalaby and Plunkett, "Standing on the Shoulders."

52. Rajeev RK (from PicoNETS), interview by Rahul Mukherjee, August 11, 2022, Mumbai.

53. Ayushi Kar, "Net Neutrality Debate Resurface as Telcos Clash with Big Tech on Regulating Content Data Networks," *Hindu Business Line*, March 7, 2022, https://www.thehindubusinessline.com/info-tech/net-neutrality-debate-resurface-as-telcos-clash-with-big-tech-on-regulating-content-data-networks/article65198440.ece.

54. Lobato, *Netflix Nations.*

55. Kar, "Net Neutrality Debate Resurface."

56. Some medium-sized ISPs these local CDNs have worked with have a maximum of 500,000 subscribers.

57. See Akamai promotions quote: Akamai, "What Is a CDN (Content Delivery Network)?" https://www.akamai.com/our-thinking/cdn/what-is-a-cdn

58. Rajeev RK, interview.

59. Prakash Advani, interview.

60. I had met Rajeev RK and Prakash Advani together at a Starbucks café in Andheri, Mumbai on August 11, 2022. It was raining heavily that day, and to save me the trouble of going from Khar West, where I was staying, to Kandivali, where the picoNETS office was, they both kindly agreed to meet me halfway in Andheri. On February 4, 2025, over an email, Prakash informed me that Rajeev had left picoNETS this year.

61. Rajeev RK, interview.

62. Netflix, "A Cooperative Approach to Content Delivery: Open Connect Briefing Paper," 2021, https://openconnect.netflix.com/Open-Connect-Briefing-Paper.pdf.

63. Rajeev RK, interview.

64. While growing up in the small city of Ranchi in late 1990s, I would call up my local cable operator in the South Office Para neighborhood after getting back from school in the afternoon. The call would be a request to play a just-released popular Hollywood film on one of the cable channels. The operator would sometimes oblige if he had the video cassette to play on the VCR. Some of these cable operators had owned video parlors earlier. At times, operators have taken some channels off the air within a particular region to regulate content based on the diktats of local party workers. For more on this, see Mishra, "Dish Is Life"; and Padma Chirumamilla, "Producing TV(s): The Multitudinous Life of Television in South India," PhD diss., University of Michigan, 2019.

65. Mishra "Dish Is Life." Some ISPs may be franchises of larger multisystem operators (MSOs) or telcos. The cable operators transitioning to ISPs are mostly those who instead of earlier providing the last mile of coaxial cable are now providing optical fiber connection. These local ISPs provide broadband Wi-Fi and don't provide any cellular internet. Several cell phone users along with TV audiences of each household then use this broadband Wi-Fi in their apartment homes. Suyash Barve writes that in dense urban locales of Mumbai, even with the rise of digital TV providers like Hathway and telcos such as Jio, cable wallas have been able to negotiate franchise deals with bigger companies and remain the last mile operators. These cable wallas stress their presence at neighborhood nakas (street intersections) and their *seva*

(service) as key to local customer satisfaction, thereby vernacularizing globally floating neoliberal ideas of enterprise. See Suyash Barve,"Enterprise Culture, neoliberalization, and cable television at the last-mile: A case-study of local cable operators in Mumbai," *Geoforum*, 150, 2024.

66. Jennifer Hessler, "To Each Their Own Ad: Nielsen and the Addressable Future of Linear TV," *Flow* 6, no. 2 (2019).

67. Chirumamilla, "Producing TV(s)."

68. Tupur Chatterjee, "The House Cannot Be Full: Risk, Anxiety, and the Politics of Collective Spectatorship in a Pandemic," *International Journal of Cultural Studies* 25, no. 3–4 (2022): 384–403.

69. Tejaswini Ganti, "Risky Business: The Structure and Practice of Formal Film Distribution in the Hindi Film Industry," in *A Companion to Indian Cinema*, ed. Neepa Majumdar and Ranjani Mazumdar (New Jersey: Wiley Blackwell, 2022).

70. Ishita Tiwary, "'OTT Is Exactly What TV Is Not': Structural Adjustment and Shifts in Indian Scriptwriting," in *Streaming Video: Storytelling Across Borders*, ed. Amanda Lotz and Ramon Lobato (New York: NYU Press, 2023).

71. Neta Alexander, "Rage Against the Machine: Buffering, Noise, and Perpetual Anxiety in the Age of Connected Viewing," *Cinema Journal* 56, no. 2 (2017): 1–24.

72. K. Sunil Thomas, "Micro Dramas Are Set To Be the Next Big Thing in Indian Entertainment," *The Week*, February 21, 2025, https://www.theweek.in/news/biz-tech/2025/02/21/micro-dramas-are-set-to-be-the-next-big-thing-in-indian-entertainment.html.

73. Tori Omega Arthur, "iRokoTV: Drama for the 'Small-Small' Screen," in *From Networks to Netflix: A Guide to Changing Channels*, 2nd ed., ed. Derek Johnson (London: Routledge, 2022), 285–286.

74. Chalaby and Plunkett, "Standing on the Shoulders," 3217.

75. Danish Khan, interview.

76. The terms "fog computing" and "mist computing" are quite loosely used in the industry. According to some local CDN players, from a technical perspective, fog computing is nothing more than a marketing term (coined by Cisco) that essentially means the same as edge computing, basically talking of fog as clouds that are close to the ground. Mist computing as a further extension of fog computing, also referred to as deep/extreme edge computing, does have the potential of bringing computing to the actual endpoint devices (not just closer to the user but all the way to the actual user). This could mean an in-home set top box that also has some local storage that can be used as a local cache.

77. Chirumamilla, "Producing TV(s)."

78. Chatterjee, "The House Cannot Be Full."

79. Vibushan Shah Subba, "The Bad-Shahs of Small Budget: The Small-Budget Hindi Film of the B Circuit," *BioScope* 7, no. 2 (2016): 215–233.

Chapter 3

1. Pankaj Mishra, interview by Rahul Mukherjee, July 2020.

2. Idrees Ali, "Social Media Played Big Role in India's Election," *Voice of America*, June 6, 2014, https://www.voanews.com/a/social-media-emerges-as-a-key-tool-in-indias-election/1931238.html.

3. Amit Bhardwaj, "How BJP's IT Cell Waged War and Won in UP," *Newslaundry*, March 17, 2017, https://www.newslaundry.com/2017/03/17/how-bjps-it-cell-waged-war-and-won-in-up.

4. Jack L. Qiu, "Three Approaches to Platform Studies: Cobweb, Billiard Balls, and Ant Societies," *Social Media + Society*, July–September 2023: 1–13.

5. Courtney Brannon Donoghue et al., "Introduction: Media Distribution Today," in *Digital Media Distribution: Portals, Platforms, Pipelines*, ed. Paul McDonald et al. (New York: NYU Press, 2021).

6. In media industry scholarship, more recently, there have been calls to think beyond the portal-platform and professional-generated content and user-generated content divides to study possible overlaps and blurring of boundaries in digital production and distribution cultures. See Smith Mehta and Stuart Cunningham, "Digitalization and Informality in Media Industries: Beyond the Platform-Portal Divide?" *Contemporary South Asia* 31, no. 2 (2023): 276–290.

7. Jose Van Dijck, *The Culture of Connectivity: A Critical History of Social Media* (Oxford: Oxford University Press, 2013), 18.

8. Daniel Miller, Elisabetta Costa, Nell Haynes, et al., *How the World Changed Social Media* (London: UCL Press), 2016.

9. Umang Bedi, "'In conversation with Umang Bedi,' Co-Founder, Verse Innovation and Dailyhunt," *India Global Forum*, December 14, 2021. https://www.youtube.com/watch?v=__1k4QMMRUU.

10. The app ecosystem in India has been crowded and cluttered, particularly since the significant rise of appified smartphones in 2016. As discussed in the other chapters, much changed with Reliance Jio's introduction of cheap data plans in 2016 and how other telecom players followed that up to remain competitive.

11. Jean Burgess, "Hearing Ordinary Voices: Cultural Studies, Vernacular Creativity and Digital Storytelling," *Continuum* 20 no. 2 (2006): 201–214; Suruchi Mazumdar, "Loving the Enemy App: Resistance vs. Professionalism in 'post-TikTok' India," *Global Media and China* 7 no.3 (2022): 340–356.

12. Timothy Havens, Amanda Lotz, and Serra Tinic. 2009. "Critical Media Industry Studies: A Research Approach." *Communication, Culture & Critique* 2, no. 2 (2009): 234–253.

13. Havens et al., "Critical Media Industry Studies," 237

14. Henry Jenkins Sam Ford, and Joshua Green, *Spreadable Media: Creating Value and Meaning in a Networked Culture* (New York: NYU Press, 2013).

15. Limor Shifman, *Memes in Digital Culture* (Cambridge, MA: MIT Press, 2013), 56.

16. Memes are also part of a group of texts, but there can indeed be one viral video that is being retransmitted without change. Of course, viral videos can be, and are indeed, altered at some stage in their circulations, so Shifman is careful to note that the viral and the memetic should be considered "two ends of a dynamic spectrum rather than as a binary dichotomy." See Shifman, *Memes in Digital Culture*, 56–57.

17. Richard Dawkins, *The Selfish Gene* (Oxford: Oxford University Press, 2016).

18. S. V. Srinivas, "Politics in the Age of YouTube: Degraded Images and Small-Screen Revolutions," in *Asian Video Cultures: In the Penumbra of the Global*, ed. Joshua Neves and Bhaskar Sarkar (Durham, NC: Duke University Press, 2017).

19. Tony Sampson, *Virality: Contagion Theory in the Age of Networks* (Minneapolis: University of Minnesota Press, 2012). Also see Gabriel Tarde, *The Laws of Imitation*, translated by E. Parsons (New York: Henry Holt, 1903).

20. Sampson, *Virality*.

21. Sampson's Tardian affective contagion theory also eschews the rigid and instrumental notion of networks with edges and nodes and proposes a more ambitious epidemiological diagram for the spread of affective contagion. Affect spreads within an affective atmosphere of encounters that is full of not just fear and anxiety but also passions, obsessions, love, hate, and other empathic transfers.

22. Furthermore, the notion that affect is both the content and the medium that circulates—that is, affect is what builds the affective atmosphere and is what also circulates within such an environment—means we are never outside affect. This seems fine for Sampson's framework, but I am left with the feeling that the affective contagion theory seems to be precise about the connection points between the social and the biological but remarkably abstract about how ordinary people's everyday lives are shaped by and shape political structures.

23. Ravi Sundaram, "Post-Postcolonial Sensory Infrastructure," *e-flux* 64 (2015), http://www.e-flux.com/journal/64/60858/post-postcolonial-sensory-infrastructure/; Sahana Udupa, "Viral Video: Mobile Media, Riot and Religious Politics," in *Media as Politics in South Asia*, ed. Sahana Udupa and Stephen D. McDowell (London: Routledge, 2017), 190–205.

24. Sangeet Kumar, "Contagious Memes, Viral Videos and Subversive Parody: The Grammar of Contention on the Indian Web," *International Communication Gazette* 77, no. 3 (2015): 232–247.

25. Aswin Punathambekar and Sriram Mohan, "A Sound Bridge: Listening for the Political in the Digital Age," *International Journal of Communication* 11 (2017): 4610–4629.

26. Anirban Baishya, "Chronicles of a Meme Foretold: Political Memes as Folk Memory in India," *Communication, Culture & Critique* 14, no. 3 (2021): 497–506.

27. Ravi Sundaram, "The Pharmacologies of Short Video, TikTok, and Beyond," in *Technopharmacology*, ed. Joshua Neves, Aleena Chia, Susanna Paasonen, and Ravi Sundaram (Minneapolis: University of Minnesota Press, 2022), 29–54.

28. Sundaram, "Pharmacologies of Short Video," 34.

29. Udupa, "Viral Video."

30. Udupa, "Viral Video"; Fathima Nizaruddin, "Role of Public WhatsApp Groups Within the Hindutva Ecosystem of Hate and Narratives of 'CoronaJihad,'" *International Journal of Communication*, 15 no. 18 (2021): 1102–1119. Also see Shakuntala Banaji and Ram Bhat "WhatsApp Vigilantes: An Exploration of Citizen Reception and Circulation of WhatsApp Information Linked to Mob Violence in India," London School of Economics and Political Science, 2019, http://www.lse.ac.uk/media-and-communications/assets/documents/research/projects/WhatsApp-Misinformation-Report.pdf.

31. Some parts of this section appeared in an essay I wrote about WhatsApp; see Rahul Mukherjee, "Mobile Witnessing on WhatsApp: Vigilante Virality and the Anatomy of Mob Lynching," *South Asian Popular Culture* 18, no. 1 (2020): 79–101.

32. Snigdha Poonam writes, "The Hindu cowboy accords to the cow the holiest status in his imagination: of mother," The cow (*gau*) is mother (*mata*), and hence, *gau-mata* (cow as mother) needs sons of the soil, the *gau-rakshaks*, to protect her. Snigdha Poonam, "The Cult of the Hindu Cowboy," *Granta*, December 9, 2016, https://granta.com/cult-hindu-cowboy. "Hindutva," mobilized in the 1920s by Vinayak Savarkar is a religio-political ideology that wants to reconfigure "India" as a "Hindu nation" where the essence of India (and being Indian) is a Hindu way of life. See Sumit Sarkar, "Indian Nationalism and the Politics of Hindutva," in *Contesting the Nation: Religion Community and the Politics of Democracy in India*, ed. D. Ludden (Philadelphia: University of Pennsylvania Press, 1996), 270–294.

33. Siddharth Narrain, "Dangerous Speech in Real Time: Social Media, Policing, and Communal Violence," *Economic and Political Weekly* 50, no. 34 (2017), https://www.epw.in/engage/article/dangerous-speech-real-time-social-media-policing-and-communal-violence. This is not to suggest that Twitter is a very open system.

Internet Hindus trolling on Twitter have used the shield of partial anonymity by using false names, aliases, nicknames, and confusing avatars. See Sriram Mohan, "Locating the 'Internet Hindu': Political Speech and Performance in Indian Cyberspace," *Television and New Media* 16, no. 4 (2015): 339–345.

34. "The Making of Lynchistan: Inside India's Deadly 'Gau Raksha' Network," *Quint*, November 20, 2018, www.youtube.com/watch?v=UaCtvYtV2f0.

35. Bajrang Dal is one of the affiliates of Rashtriya Swayamsevak Sangh (RSS), the principal conservative Hindu "cultural" organization that is connected with the political party Bharatiya Janata Party (BJP).

36. Betwa Sharma, "Despite Horrific Killing over Beef, Communal Hatred Is Reluctant to Take Root in Bisada Village," HuffPost, October 1, 2015, www.huffingtonpost.in/2015/10/01/how-centuries-of-peace-in_n_8224590.html.

37. Supriya Sharma, "As Meat Is Turned into a Political Issue in India, a Muslim Ironsmith in Dadri Pays the Price," Scroll.in, October 2, 2015, https://scroll.in/article/759319/as-meat-is-turned-%20into-a-political-issue-across-india-a-muslim-ironsmith-in-dadri-pays-the-price.

38. Sudha Pai and Sajjan Kumar, *Everyday Communalism: Riots in Contemporary Uttar Pradesh* (New Delhi: Oxford University Press India, 2018).

39. Mohammad Ali, interview by Rahul Mukherjee, June 4, 2018.

40. Sharma, "As Meat Is Turned."

41. Paul R. Brass, *The Production of Hindu-Muslim Violence in Contemporary India* (New Delhi: Oxford University Press, 2003).

42. Fathima Nizaruddin, "Institutionalized Riot Networks in India and Mobile Instant Messaging Platforms," *Asiascape: Digital Asia* 9, no. 1–2 (2022): 71–94.

43. Wendy Chun, *Updating to Remain the Same: Habitual New Media* (Cambridge, MA: MIT Press, 2016). Affect theorist Carolyn Pedwell writes, "Affect may offer the spark that catalyzes embodied transformation, yet without some form of habituation, enduring change may fail to take shape." Carolyn Pedwell, "Mediated Habits: Images, Networked Affect and Social Change." *Subjectivity* 10, no. 2 (2017): 147–169. "Affective life," Margaret Wetherell has argued, is not only about "spontaneous, extraordinary, one-off activities" but is also modulated through longer processes of emotional and daily patterns. Margaret Wetherell, *Affect and Emotion: A New Social Science Understanding* (London: Sage, 2012).

44. Radhika Govindrajan, "Labors of Love: On the Political Economies and Ethics of Bovine Politics in Himalayan India," *Cultural Anthropology* 36, no. 2 (2021): 193–221. While it is important to retain the difference between *gau raksha* (cow protection) and *gau seva* (cow service), there is a blurring between the two forms of actions on the ground.

45. Banaji and Bhat, "WhatsApp Vigilantes."

46. Arjun Appadurai, "A Syndrome of Aspirational Hatred Is Pervading India," Wire .in, December 10, 2019, https://thewire.in/politics/unnao-citizenship-bill-violence -india.

47. Furthermore, Snigdha Poonam notes that such is the religiopolitical climate today in India that this recognition, which upper-caste Hindu men earlier did not need to ask for, today has to be actively sought by them through such deplorable acts. Snigdha Poonam, *Dreamers: How Young Indians Are Changing the World* (Cambridge, MA: Harvard University Press, 2018).

48. Chacko, "Gender and Authoritarian Populism"

49. Banaji and Bhat, "WhatsApp Vigilantes."

50. Nikhil Pahwa, "Addressing WhatsApp's Fake News Problems," *Medianama*, July 3, 2018, https://www.medianama.com/2018/07/223-addressing-whatsapps-fake -news-problems/.

51. Gowhar Farooq, "Old Conflicts in New Media Assemblages: India's Cow Vigilantism and YouTube," *Asiascape: Digital Asia* 9, no. 1–2 (2022): 47–70.

52. Kunal Purohit, *H-Pop: The Secretive World of Hindutva Pop Stars* (Gurugram: HarperCollins, 2023).

53. Purohit, *H-Pop*, xix.

54. Radhika Gajjala and Tarishi Verma, "WhatsAppified Diasporas and Transnational Circuits of Affect and Relationality," in *Applied: Culture in the Age of Apps*, ed. Jeremy Wade Morris and Sarah Murray (Ann Arbor: University of Michigan Press, 2018), 205–215. See Debjani Chakraborty, "Domesticating technology: Learning from the use of mobile phone of small women rural entrepreneurs in India during COVID-19," *Mobile Media & Communication* 12, no. 3 (2024): 499–516.

55. Amrita Chakravarty, "Film(i) Culture After Film: Lip-sync Media and the Expanded Archive of Hindi Cinema," *BioScope* 14, no. 2 (2023): 186–209.

56. Diana Zulli and David James Zulli, "Extending the Internet Meme: Conceptualizing Technological Mimesis and Imitation Publics on the TikTok Platform," *New Media and Society* 24, no. 8 (2022): 1872–1890.

57. "Har Ghoont Mein Swag," https://www.youtube.com/watch?v=I5cPOd9aCUc.

58. Adgully Bureau, "Pepsi Swag Anthem Creates Record on TikTok, *Adgully*, May 6, 2019, https://www.adgully.com/pepsi-s-swag-anthem-creates-new-record-on-tiktok -85439.html.

59. "Swagwala Step Challenge TikTok," https://www.youtube.com/watch?v=wEGuP DxhU3I.

60. Neha Dewan, "On TikTok, Everyone's Up for a Challenge," *Economic Times*, August 6, 2019, https://economictimes.indiatimes.com/small-biz/marketing-branding/marketing/on-tiktok-everyones-up-for-a-challenge/articleshow/70459209.cms?from=mdr.

61. Yashraj Sharma, "Instagram Has Largely Replaced TikTok in India, and Erased Working-Class Creators," *Rest of World*, October 6, 2021, https://restofworld.org/2021/instagram-and-class-in-india/.

62. Nilesh Christopher and Varsha Bansal, "Meet the Rising Stars of India's Video Apps," *Rest of World*, May 12, 2020, https://restofworld.org/2020/india-video-apps-influencers/.

63. Min Zhou and Shih Diing Liu, "Becoming Precarious Playbour: Chinese Migrant Youth on the *Kuaishou* Video- Sharing Platform," *Economic and Labor Relations Review* 32, no. 3 (2021): 322–340. Miao Li, Chris KK Tan, and Yuting Yang, "*Shehui Ren*: Cultural Production and Rural Youths' Use of the *Kuaishou* Video-Sharing App in Eastern China," *Information, Communication & Society* 23, no. 10 (2019): 1499–1514.

64. Nilesh Christopher, "TikTok Is Fueling India's Deadly Hate Speech Epidemic," *Wired*, August 12, 2019, https://www.wired.co.uk/article/tiktok-india-hate-speech-caste.

65. Nilesh Christopher, interview by Rahul Mukherjee, June 26, 2020.

66. Christopher, "TikTok Is Fueling."

67. Nilesh Christopher, interview.

68. Shadma Shaikh, "Chinese Apps Scramble for Indian Kardashians," *Factor-Daily*, March 27, 2019, https://factordaily.com/chinese-apps-scrambling-for-indias-kardashians/.

69. Shaikh, "Chinese Apps Scramble."

70. Shadma Shaikh, interview by Rahul Mukherjee, August 4, 2019.

71. Zulli and Zulli, "Extending the Internet Meme."

72. Munsif Vengattil, "The Cracks in ShareChat, Moj's Content Moderation Machine," *The Ken*, May 17, 2021, https://the-ken.com/story/the-cracks-in-sharechat-mojs-content-moderation-machine/.

73. Shrabonti Bagchi, "Sharechat: The Indian Language Social Network That Gets Bharat in a Way Facebook Doesn't," Factordaily, October 24, 2016, https://archive.factordaily.com/indian-language-social-network-gets-bharat-way-facebook-cannot/.

74. This small and snackable kind of content approach worked well to create vernacular sociality in India with templates for good-morning messages, religion-related stickers, messages about fans of Bollywood megastars Shahrukh and Salman Khan,

and messages related to the six 6s that cricketer Yuvraj Singh struck. These Sharechat messages in regional languages were being shared through WhatsApp as well. This is how Sharechat took off as one of the earliest players in the Indian regional content social media space. For more, refer to Bagchi, "Sharechat."

75. Journalist, interview by Rahul Mukherjee, December 2022.

76. Raghavendra Prajapati, interview.

77. Kyle S. Barnett, "Record Men: Talent Scouts in the U.S. Record-ing Industry, 1920-1935," In Derek Jhonson, Derek Kompare, and Avi Santo (eds.), *Making Media Work: Cultures of Management in the Entertainment Industries* (New York: NYU Press, 2014), 113–142.

78. Mukherjee, *Aspirational Politics of Talent Acquisition.*

79. Raghavendra Prajapati, interview.

80. Barnett, "Record Men,"114.

81. The short-video platform sector has witnessed some consolidation with Mohalla Tech, Moj's parent company, acquiring MX TakaTak, which was previously owned by the Times Internet. This merger made Moj the largest short video platform in the country. Josh has remained competitive with Moj. Josh was founded by Verse Innovation, which also runs the news aggregator site, Dailyhunt. See Naman Ramachandran, "India's Moj, MX TakaTak Merge to form Largest Short Video Platform in the Country, Where TikTok Remains Banned," *Variety*, February 10, 2022, https://variety.com/2022/digital/news/moj-mx-takatak-merger-short-video-platform-1235177763/.

82. Seher Bedi, interview by Rahul Mukherjee, January 10, 2023.

83. See Mukherjee, *Aspirational Politics of Talent Acquisition*

84. The award ceremony was hosted by Bollywood actors Vivek Oberoi and Urvashi Rautela with celebrity influencer Faisu. While handing out awards, Bollywood celebrities acknowledged the power and influence of short video content in brief speeches with actor Suniel Shetty and choreographer and director Farah Khan, remarking about just how much money an influencer and content creator like Faisu was making.

85. Alice E. Marwick, "Instafame: Luxury Selfies in the Attention Economy," *Public Culture* 27 (2015): 137–160.

86. Content director, interview by Rahul Mukherjee, December 21, 2022.

87. Seher Bedi, interview.

88. Rosie Findlay, "Trust Us, We're You": Aspirational Realness in the Digital Communication of Contemporary Fashion and Beauty Brands," *Communication, Culture,*

& *Critique* 12, no. 4 (2019): 553–569; Brooke Erin Duffy, *Not Getting Paid To Do What You Love: Gender, Social Media, and Aspirational Work* (New Haven, CT: Yale University Press, 2017).

89. Taina Bucher and Anne Helmond, "The Affordances of Social Media Platforms," in *The SAGE Handbook of Social Media*, ed. Jean Burgess, Alice Marwick, and Thomas Poell (London: Sage, 2018), 233–253.

90. Smith Mehta and D. Bondy Valdovinos Kaye, "Pushing the Next Level: Investigating Digital Content Creation in India," *Television & New Media* 22, no. 4 (2022): 360–378.

91. Joshua Braun, "Points of Origin: Asking Questions in Distribution Research," in *Digital Media Distribution: Portals, Platforms, Pipelines*, ed. Courtney Brannon Donoghue et al. (New York: NYU Press, 2021), 32.

92. Nitin Govil and Anirban K. Baishya, "The Bully in the Pulpit: Autocracy, Digital Social Media, and Right-wing Populist Technoculture," *Communication Culture & Critique* 11, no. 1 (2018): 67–84.

93. Sharma, "Instagram Has Largely Replaced." Of course, it remains to be said that there are many different kinds of content creators in such a diverse country as India based on differences of specialization/occupational sector (fitness, make-up, self-help), regional differences, and differences of caste, gender, and class. Furthermore, there are creators who can move from being a microinfluencer to an influencer or climb up the professional aspirational ladder from amateur creator or fan content maker to professional insider; see Mehta and Kaye, "Pushing the Next Level."

94. Outreach efforts to tap local talent in various linguistically diverse regions of India are not a completely new initiative of the contemporary short-video platforms. In the past, YouTube and Facebook also organized workshops and other events to facilitate engagement and collaboration between regional advertisers, content creators, and users. See Smith Mehta, "Localization, Diversification and Heterogeneity: Understanding the Linguistic and Cultural Logics of Indian New Media," *International Journal of Cultural Studies* 23, no. 1 (2020): 102–120. In its initial years, starting from 2008, YouTube provided a platform to independent artists, but over the years it became institutionalized to the extent that today independent performers are sometimes unable to directly approach YouTube but need to associate themselves with a multichannel network (MCN) to negotiate with YouTube. That said, given its early lead in the streaming video revolution, YouTube continues to be popular across all segments of Indian society and among all tiers of cities. Anaar Desai-Stephens, "The Infrastructure of Engagement: Musical Aesthetics and the Rise of YouTube in India," *Twentieth Century Music* 19, no. 3 (2022): 444–471.

95. Shudeep Majumdar, interview excerpt, January 11, 2023.

96. Harveen Ahluwalia, "The Surprising Bad News at Dailyhunt and Josh," *Morning Context,* December 2, 2022, https://themorningcontext.com/internet/the-surprising-bad-news-at-dailyhunt-and-josh.

97. Marwick, "Instafame."

98. Jamie Coates, "So 'Hot' Right Now: Reflections on Virality and Sociality from Transnational Digital China," *Digital Culture & Society, Mobile Digital Practices* 3, no. 2 (2017): 91.

99. Shrishti Jaswal, "Indian Politicians Embrace Influencers Ahead of 2024 Elections," *Rest of World,* July 24, 2023, https://restofworld.org/2023/india-2024-elections-influencers/.

100. Joyojeet Pal, "Journalism and the Weight of Social Media Influence," *Media Runble,* December 19, 2023, https://www.themediarumble.com/event/social-media-influencers-and-political-capture-in-india

101. See Shailesh's Instagram account for the videos: https://www.instagram.com/shailesh_ki_duniya/.

102. AI deepfake technologies re-created a politician's voice and video, and phone audiences in the 2024 Indian national elections received video messages in which they could see their political representatives address them by name and talk with eerie particularity about the local issues that matter most to them. This was not the candidate speaking, but their deepfake. With such personalized messages, campaigners hoped that voters/audiences would vote for a particular political party's candidate, who in some cases did not have the time to knock on doors or exchange a greeting or handshake. See Nilesh Christopher and Varsha Bansal, "Indian Voters Are Being Bombarded with Millions of Deepfakes. Political Candidates Approve," *WIRED*, May 28, 2024, https://www.wired.com/story/indian-elections-ai-deepfakes/.

Chapter 4

1. Digital India, *Ministry of Electronics and Information Technology, Government of India*, 2017, http://www.digitalindia.gov.in/.

2. Ajit Ranade, "Role of 'Fintech' in Financial Inclusion and New Business Models," *Economic & Political Weekly* 52, no. 12 (2017): 125–128.

3. Adrian Athique, "A Great Leap of Faith: The Cashless Agenda in Digital India," *New Media & Society* 21, no. 8 (2019): 1697–1713.

4. For those who do not have either a smartphone or a debit card, there is the Aadhaar-enabled payment system so that they can withdraw or transfer money using their fingerprint. Essentially, this is all done to ensure that everybody has a bank account.

5. Vijayanka Nair, "Governing India in Cybertime: Biometric IDs, Start-Ups, and the Temporalized State," *South Asia: Journal of South Asian Studies* 42, no. 3 (2019): 520.

6. Athique, "A Great Leap of Faith." Also see Ajay Gandhi, "Black Money in India: Fighting Specters and Fostering Relations," in *Rethinking Markets in Modern India: Embedded Exchange and Contested Jurisdiction*, ed. Ajay Gandhi, Barbara Harriss-White, Douglas Haynes, and Sebastian Schwecke (Cambridge, UK: Cambridge University Press, 2020), 269–293.

7. Aditi Bhatia-Kalluri and Brett Caraway, "Transformation of the Digital Payment Ecosystem in India: A Case Study of Paytm," *Social Inclusion* 11, no. 3 (2023): 1–12.

8. Sanjay Phadke, *FinTech Future: The Digital DNA of Finance* (New Delhi: Sage Publications India, 2020).

9. Nair, "Governing India in Cybertime."

10. Since the Netflix TV series, several short documentaries, including one by Netflix called *Jamtara's Credit Card Scammers* (Netflix India, 2020, 6 mins) have been made. On a similar topic, NDTV made a documentary about OTP (one-time password) scams in Nuh in Haryana, where similar phishing operations were being carried out, titled *OTP Mafia* (2023). CNA Insider made a documentary about tech support scams in India and the role of the call center industry, *India's Thriving Tech Support: Before You Call Tech Support* (2022). While I do occasionally refer to some of these documentaries, I mostly focus on the Netflix show.

11. Jovan Scott Lewis, *Scammer's Yard: The Crime of Black Repair in Jamaica* (Minneapolis: University of Minnesota Press, 2020); Brian Larkin, *Signal and Noise: Media, Infrastructure, and Urban Culture in Nigeria* (Durham, NC: Duke University Press, 2008).

12. Lewis, "Scammer's Yard."

13. Winifred Poster, "Outsourcing, the Gig Economy, and International Scam," *Symposium—Gig Economy: University of Pacific Law Review* 54, no. 1 (2023): 54–65.

14. Prem Mahadevan, "A Social Anthropology of Cybercrime: The Digitization of India's Economic Periphery," April 2020, https://globalinitiative.net/wp-content/uploads/2020/04/India-Cybercrime.10.04.web_.pdf.

15. Bagish Jha, "Cyber Fraud: How a New 'Jamtara' Has Emerged at Tri-junction of States," *Times of India*, July 9, 2020, http://timesofindia.indiatimes.com/articleshow/76863779.cms?utm_source=contentofinterest&utm_medium=text&utm_campaign=cppst.

16. Mary Ann Joley and David Boyle, "Forced to Scam: Cambodia's Cyber Slaves," *Al-Jazeera*, August 11, 2022, https://www.aljazeera.com/features/longform/2022/8/11/meet-cambodia-cyber-slaves.

17. Aneesh Aneesh, *Virtual Migration: The Programming of Globalization* (Durham, NC: Duke University Press, 2006). Also see Lewis, "Scammer's Yard."

18. Lewis, "Scammer's Yard," 9.

19. Poster, "Outsourcing, the Gig Economy." Also see Enda Brophy and Rodrigo Finkelstein, "Digital Debt Labour: Migration, Deportation and Offshoring in Mexico," *tripleC* 18, no. 1 (2020): 304–321.

20. Poster, "Outsourcing, the Gig Economy," 54.

21. Poster, "Outsourcing, the Gig Economy," 57.

22. Yudhijit Bhattacharjee, "Who's Making All These Scam Calls?" *New York Times*, January 27, 2021, https://www.nytimes.com/2021/01/27/magazine/scam-call-centers.html; Poster, "Outsourcing, the Gig Economy."

23. Winifred Poster, "Sound Bites, Sentiments, and Accents: Digitizing Communicative Labor in the Era of Global Outsourcing," in *digitalSTS: A Field Guide for Science and Technology Studies*, ed. Janet Vertesi and David Ribes (Princeton, NJ: Princeton University Press, 2019), 240.

24. Cited in Deepu Sebastian Edmond, "Phishing in Jamtara: What Does It Take to Carry Out Online Fraud?" *Indian Express*, December 13, 2015, https://indianexpress.com/article/india/india-news- india/phishing-in-jamtara-what-does-it-take-to-carry-out-online-fraud/.

25. Aparna Nandakumar, "Rebels for Every Cause: Mythologies of Youth in Contemporary India," *Studies in South Asian Film & Media* 11, no. 1 (2020): 87–100.

26. Pankaj Mishra, "Jamtara, Where India Is Winning Its War on Cyber Crime," *Factor Daily*, May 7, 2018, https://archive.factordaily.com/india-winning-war-on-cyber-crime-jamtara/.

27. Kentaro Toyama, ed., "Review of Research on Rural PC Kiosks. Information Technology and Communications Resources for Sustainable Development," *EOLSS*, 2004, https://www.eolss.net/sample-chapters/c15/e1-25-02-04.pdf.

28. Ajay Kumar, "Money Siphoned in Cyber Fraud Finds Way to Jan Dhan Accounts: An Investigation," *India Today*, December 19, 2019, https://www.indiatoday.in/top-stories/story/money-siphoned-in-cyber-fraud-finds-way-to-jan-dhan-accounts-an-investigation-1629509-2019-12-19; Kalyan Das, "Jan Dhan Yojana Accounts on Radar of Cyber Criminals," *Hindustan Times*, August 30, 2016, https://www.hindustantimes.com/bhopal/jan-dhan-yojana-accounts-on-radar-of-cyber-criminals/story-gW7aJn3xXmXPGoiwUa4uSK.html.

29. "ED Attaches Assets of Jamtara-Based Cyber Criminals," *Economic Times*, March 11, 2021, https://economictimes.indiatimes.com/news/politics-and-nation/ed-attaches-assets-of-jamtara-based-cyber-criminals/articleshow/81444906.cms.

30. Anumeha Yadav, "JAM in Jharkhand: 'Apply Lemon Juice, Flour, Boroplus on Fingers and Pass Biometrics Test,'" *Scroll.in*, November 23, 2015, https://scroll.in/article/769611/jam-in-jharkhand-apply-lemon-juice-flour-boroplus-on-fingers-and-pass-biometrics-test; Anumeha Yadav, "Jan Dhan Yojana: On Paper, a Radical Scheme. On the Ground, a Catalyst for Confusion and Coercion," *Scroll.in*, November 24, 2015, https://scroll.in/article/769613/jan-dhan-yojana-on-paper-a-radical-scheme-on-the-ground-a-catalyst-for-confusion.

31. See Yadav, "JAM in Jharkhand"; Yadav "Jan Dhan Yojna." Here, I will not get into the interfacing problems of various Aadhaar and allied media technologies, whether it is the iris reader or the biometric scanner, which sometimes simply does not recognize fingerprints. Ursula Rao, "Biometric Marginality: UID and the Shaping of Homeless Identities in the City," *Economic & Political Weekly* 48, no. 13 (2013): 72–77. The government characterizes such problems as teething troubles to be resolved soon. It also remains highly debatable whether Aadhaar has made it easier for the poor in India to access welfare benefits or has created more barriers to receiving such benefits. See Itty Abraham and Asish Rajadhyaksha, "State Power and Technological Citizenship in India: From the Postcolonial to the Digital Age," in *East Asian Science, Technology and Society: An International Journal* 9, no. 1 (2014): 65–85. In any case, the Aadhaar mode of biopolitical control remains as much about control (and surveillance of citizen data) as it is about care (and welfare distribution). See Nafis Aziz Hasan, "Bureaucratic Mediations for Biometric Governance in India's Northeast—Aadhaar in Tripura," *South Asia: Journal of South Asian Studies*, March 5 2022, https://www.tandfonline.com/doi/full/10.1080/00856401.2022.2035976.

32. Chacko, "Gender and Authoritarian Populism."

33. Chacko, "Gender and Authoritarian Populism," 210. According to Chacko, these schemes targeting women position them as "responsible, sacrificing, laboring" and Modi as a "transformational, benevolent patriarch," and thus the construction of the "virtuous market citizen" serves both neoliberal governmentality and Hindutva Nationalism.

34. Christophe Jaffrelot, *Modi's India: Hindu Nationalism and the Rise of Ethnic Democracy* (Princeton, NJ: Princeton University Press, 2021).

35. Jaffrelot, "Modi's India," 120.

36. Jamtara villager, interview by Rahul Mukherjee, July 2021.

37. The use of the colloquial term *dhanda* for business is significant. "Dhanda" as Anastasia Piliavsky notes, is an activity that has a vital presence in the everyday lives of aspiring rural youths. It is both elusive and an all-pervading preoccupation; you might think people are idling or chatting away at tea stalls and the village *hath* (bazaar), but they are actually engaging in *dhanda*. Depending on context, *dhanda* could mean simply "business" or "political entrepreneurship" or "sex-work."

Anastasia Piliavsky, "Against the Public Sphere: The Morals of Disclosure and the "Vernacular Public Sphere" in Rural India," in *Democratic Transformation and the Vernacular Public Arena in India*, ed. Tabarez Neyazi, Akio Tanabe, and Shinya Ishizaka (London: Routledge, 2014), 172–189.

38. Teenager in Narayanpur, Jamtara, interview by Rahul Mukherjee, July 20, 2024.

39. Teenager in Narayanpur, Jamtara, interview.

40. For the term "subaltern recalcitrance," I am grateful for conversations with David Llyod, who has written about this in the context of the Irish working class: David Llyod, "The Subaltern in Motion: Subalternity, the Popular and Irish Working Class History," *Postcolonial Studies* 8, no. 4 (2005): 421–437.

41. Jamtara local journalist, interview by Rahul Mukherjee, July 16, 2021.

42. Cybercrime police person, interview by Rahul Mukherjee, July 16, 2021.

43. Joyojeet Pal, "The Technological Self in India: From Tech-savvy Farmers to a Selfie- Tweeting Prime Minister," *ICTD '17,* November 16–19, 2017, Lahore, Pakistan; Kavita Philip, "Telling Histories of the Future: The Imaginaries of Indian Technoscience," *Identities* 23, no. 3 (2016): 276–293.

44. Snigdha Poonam, interviewed by Professor Mona Mehta as part of the Ahmedabad University session "Seminar and Lecture Series | Aspiration to Desperation," April 17, 2021, https://www.youtube.com/watch?v=OhTElk5w_kw.

45. Poonam, quoted in the "Aspiration to Desperation."

46. N., interview by Rahul Mukherjee, September 3, 2020.

47. N., interview.

48. The web series' primary protagonist is the mafia strongman character Akhilanand Tripathi (Pankaj Tripathi), who runs a covert guns and drugs operation under the guise of a carpet factory as a front in Mirzapur. "Carpet" translates as *kaleen*, and hence Tripathi's nickname is *Kaleen Bhayya* (Carpet Brother). Kaleen Bhayya is a Brahmin gangster who is the "king" of Mirzapur, and his son (and heir apparent) Munna wants to be the king of Mirzapur one day, and since he cannot be yet, he wants to be the king of the Gajjumal college.

49. These juxtapositions of dacoitry/banditry/crime suggest a way of thinking of the acts of impersonating and reassembling/reappropriating distribution infrastructures by Jamtara's scammers as part of conversations about rebellion, insurgency, and peasant unrest that have been part of the subaltern studies literature. Districts bordering the Jamtara district were part of Damin-i-koh, which included other border districts of present-day West Bengal and Jharkhand, like Dumka and Sahibganj. This was a region marked out by the British between 1780 and 1840 to separate and contain the Santals, a peripatetic tribal group, and coerce them to clear forests for

cultivation. Subaltern studies scholarship has been invested in space. Swati Chattopadhyay has written about the Santal Rebellion of 1855–1856 in relation to infrastructure (as facilitating movements across spaces) and informality and noted that the evidence from the colonial archive emphasized the "importance of infrastructure as a defining condition of the rebellion." The difficulties encountered by colonial administrators in comprehending the rebellion included the "rapid movement of the rebels, their capacity to cover extensive territory, their large gatherings, method of communication, and territorial claims." If the Santals disrupted the colonial communication system by targeting the infrastructure of colonial governance (the railways and the post), the Jamtara scammers are also subverting the logics of digital communication and mobile money distribution systems. Their scamming methods do not simply include a single con (confidence) man but a networked operation. These operations could be read as individual acts of bravado or "fugitive insurgent politics" depending on whether one finds in them stories of individual aspirations and enterprise or stories of collective resistance and world-making. See Swati Chattopadhyay, "Informality and Infrastructure," in *Urban Informality and the Built Environment: Infrastructure, Exchange, and Image*, ed. Colin Marx, Nikhilesh Sinha and Nerea Amoros (Chicago: University of Chicago Press). Also see Lewis, *Scammers Yard.*

50. Soumendra Padhi, the director of the Netflix series *Jamtara*, treats the relationship between Gudiya and Sunny with great restraint, never having any dramatized scene showing romantic expressions between the two, and yet as the show progresses, they seem to have affection for one another.

51. Purnima Mankekar, *Unsettling India: Affect, Temporality, Transnationality* (Durham, NC: Duke University Press, 2015), 190.

52. Deepu Sebastian Edmond, interview by Rahul Mukherjee, August 12, 2020.

53. Pankaj Mishra, interview by Rahul Mukherjee, July 2020.

54. That said, after demonetization, black money was not sufficiently recovered, as people invented more creative ways to hide their untaxed wealth. See Kaytie Nelsen, "How Indians Dodged Government's 'Black Money' Clampdown," Al Jazeera, February 15, 2017, https://www.aljazeera.com/features/2017/2/15/how-indians-dodged-governments-black-money-clampdown.

55. Lewis, *Scammer's Yard.*

56. For one example of such reportage on news channels, see this *India Today* NewsTrack report: "The Jamtara Files: Modus Operandi of Cyber Crime Exposed, How Can You Save Yourself?" October 12, 2022, https://www.youtube.com/watch?v=IAvk5GMc0vY. Also see NDTV's *OTP Mafia* (August 25, 2023) for stories involving mobile phone sextortion calls: https://www.youtube.com/watch?v=BSM6-gET8Vo.

57. See Priya Pathak, "WhatsApp Private Job Scams on the Rise in India: How to Stay Safe Online, All Your Questions Answered," Financial Express, May 22, 2023,

https://www.financialexpress.com/life/technology-whatsapp-private-job-scams-on-the-rise-in-india-how-to-stay-safe-online-all-your-questions-answered-3097094/. Also see the Lallantop video WhatsApp Scam: https://www.youtube.com/watch?v=caXA-Qas5ME.

58. Roundtable discussion, "Misinformation in/and Fintech" ADM+S Center, Western Sydney University, July 7, 2023.

59. An *Economic Times* article calls such work "the dark side of India's tech sector." For more, see "Gig Workers, Call Centre Scams, Ghost Work: The Dark Side of India's Tech Sector," *Economic Times*, Mar 22, 2023, https://economictimes.indiatimes.com/tech/technology/gigs-scams-ghost-work-the-dark-side-of-indias-tech-sector/articleshow/98896795.cms?utm_source=contentofinterest&utm_medium=text&utm_campaign=cppst.

60. Mary L. Gray and Siddharth Suri, *Ghost Work: How to Stop Silicon Valley from Building a New Global Underclass* (Boston: Houghton Mifflin Harcourt, 2019).

61. Sarah Roberts, *Behind the Screen: Content Moderation in the Shadows of Social Media* (New Haven, CT: Yale University Press, 2019).

62. Alastair McCready, "From Industrial-Scale Scam Centers, Trafficking Victims Are Being Forced to Steal Billions," *Vice World News*, July 13, 2022, https://www.vice.com/en/article/n7zb5d/pig-butchering-scam-cambodia-trafficking.

63. McCready, "From Industrial-Scale Scam Centers."

64. Joley and Boyle, "Forced to Scam."

65. Sandeep Sukhtankar and Milan Vaishnav, "Corruption in India: Bridging Research Evidence and Policy Options," *India Policy Forum* 11 (2015): 193–261.

66. Gandhi, "Black Money in India," 284.

67. Gandhi, "Black Money in India," 284.

68. Malini Ranganathan, David L. Pike and Sapana Doshi eds., *Corruption Plots: Stories, Ethics, and Publics of the Late Capitalist City* (Ithaca, NY: Cornell University Press, 2023). Also see Athique, "A Great Leap of Faith."

69. Sareh Pouryousefi and Jeff Frooman, "The Consumer Scam: An Agency-Theoretic Approach," *Journal of Business Ethics* 154 (2019): 1–12.

Chapter 5

1. Brett Scott, *Cloudmoney: Cash, Cards, Crypto, and the War for our Wallets* (Penguin: London, 2022).

2. Andrea Pollio and Liza Rose Cirolia, "Fintech Urbanism in the Startup Capital of Africa," *Journal of Cultural Economy* 15, no. 4 (2022): 508–523.

3. Kevin Donovan and Emma Park, "Algorithmic Intimacy: The Data Economy of Predatory Inclusion in Kenya," *Social Anthropology* 30, no. 2 (2022): 120–139.

4. Tarunima Prabhakar, "A New Era for Credit Scoring: Financial Inclusion, Data Security, and Privacy Protection in the Age of Digital Lending," UC Berkeley Center for Long Term Cybersecurity White Paper, 2020, https://cltc.berkeley.edu/wp-content/uploads/2020/06/A_New_Era_for_Credit_Scoring.pdf. Also see Donovan and Park, "Algorithmic Intimacy."

5. Rahul Mukherjee, "In/Conveniences of Mobile Payments: 'Alternative Data' and the Distribution Infrastructures of Loan Apps," Special Issue on *In/Convenience* (Institute of Network Cultures, 2024), 131–147.

6. Nilesh Christopher, "Debt and Shame via GooglePlay," *Rest of World*, January 27, 2021, https://restofworld.org/2021/debt-and-shame-via-google-play/.

7. Marc Steinberg and Joshua Neves, "Pandemic Platforms: How Convenience Shapes the Inequality of Crisis," in *Pandemic Media: Preliminary Notes Towards an Inventory* (Meson Press, 2020).

8. Lana Swartz, *New Money: How Payment Became Social Media* (New Haven, CT: Yale University Press, 2020).

9. Yichen Rao and Tom McDonald, "Debt at a Distance: Counter-Collection Strategies and Financial Subjectivities of China's Working-Class Defaulters during COVID-19," *Economy and Society* 52, no. 2 (2023): 250–273.

10. The China and India cases are also similar because those most affected by the microloan debt trap and by the pandemic-induced lockdown were working-class migratory laborers in both the countries, who found themselves without work and thus were bound to be defaulters who had no way to pay. Both countries have a history of internal migration, with laborers migrating from one part of the country to another (from agrarian regions to industrial regions or from villages to cities) in search of work.

11. Rao and McDonald, *Debt at a Distance.*

12. Winifred R. Poster, "Introduction to Special Issue on Scams, Fakes, and Frauds," *New Media & Society* 24, no. 7 (2022): 1537.

13. Lana Swartz, "Theorizing the 2017 Blockchain ICO Bubble as Network Scam," *New Media & Society* 24, no.7 (2022): 1695.

14. Marie Langevin, "Big Data for (Not So) Small Loans: Technological Infrastructures and the Massification of Fringe Finance," *Review of International Political Economy* 26, no. 5 (2019):790–814.

15. I have explained the process of intermediation involved in digital lending in much greater detail elsewhere in another essay. See Rahul Mukherjee,

"Intermediation of Lending: Platforms, Mobile Money, and Data Transactions," *Platforms & Society* 2 (2025): 1–13.

16. Paul Langley and Andrew Leyshon, "The Platform Political Economy of Fin-Tech: Reintermediation, Consolidation and Capitalisation," *New Political Economy* 26, no. 3 (2021): 376–388.

17. Arundhati Ramanathan, "The Unregulated Tech Supermarket Powering India's Loan App Scourge," *The Ken*, March 17, 2021, https://the-ken.com/story/the-unregulated-tech-supermarket-powering-loan-app/.

18. Michael Dieter, Caroline Gerlitz, Anne Helmond, et al., "Multi-Situated App Studies: Methods and Propositions," *Social Media + Society* 5, no. 2 (2019): 1–15.

19. Randy Martin, *Financialization of Daily Life* (Philadelphia: Temple University Press, 2002).

20. Léna Pellandini-Simányi Ferenc Hammer and Zsuzsanna Vargha, "The Financialization of Everyday Life or the Domestication of Finance?" *Cultural Studies* 29, no. 5–6 (2015): 733–759. Since the success of the mobile phone-based payment service M-PESA, Kenya has been the frontier of fintech experiments. Anthropologists Kevin Donovan and Emma Park discuss ways in which digital lenders capitalize on intimacy to convert kinship networks into frontiers of extraction. See Donovan and Park, "Algorithmic Intimacy." The connection between elevated aspirations and proliferation of credit sources is made by Deborah James in her discussion of the rise of "new middle class" in South Africa. It is therefore not surprising that in India a neo-middle class is rising as credit gets liberalized. The question remains how far Indians in this new class group are ready to go into debt to reach their aspirations. See Deborah James, *Money from Nothing: Indebtedness and Aspiration in South Africa* (California: Stanford University Press, 2015).

21. Kunal Tulsani, "Startup Story of Paytm: Initial Years, Growth, and Struggle," *DigitalYug*, April 20, 2021, https://thedigitalyug.com/startup/startup-story-of-paytm-initial-years-growth-and-struggle.

22. UPI launch press release by NPCI, April 11, 2016, https://www.npci.org.in/PDF/npci/press-releases/2016/UPI_Launch_Press_Release_April_11_2016.pdf.

23. iSpirt, "The India Stack Takes Digital India Campaign to a Whole New Level," https://ispirt.in/our-industry/indiastack/.

24. Ananya Roy, "Subjects of Risk: Technologies of Gender in the Making of Millennial Modernity," *Public Culture* 24, no. 1 (2012): 131–155.

25. Julia Elyachar, "Next Practices: Knowledge, Infrastructure, and Public Goods at the Bottom of the Pyramid," *Public Culture* 24, no. 1 (2012): 109–129.

26. Sohini Kar, *Financializing Poverty: Labor and Risk in Indian Microfinance* (Palo Alto, CA: Stanford University Press, 2018), 33.

27. Anthropologist Lamia Karim's research showed that Grameen Bank loans caused financial and social ruin for some women. In exchange for the loan, these women collateralized the social ties that have kept them afloat in the past. Lamia Karim, *Microfinance and Its Discontents: Women in Debt in Bangladesh* (Minneapolis: University of Minnesota Press, 2011).

28. Steve Hollingworth interviewed by Knowledge at Wharton staff, January 10, 2019, https://knowledge.wharton.upenn.edu/article/grameen-foundation-using-data-lever-combat-poverty/.

29. Ajay Kumar, "Money Siphoned in Cyber Fraud Finds Way to Jan Dhan Accounts: An Investigation," *India Today*, December 19, 2019, https://www.indiatoday.in/top stories/story/money-siphoned-in-cyber-fraud-finds-way-to-jan-dhan-accounts-an-investigation-1629509-2019-12-19.

30. Advait Palepu and Ashwin Manikandan, "Aadhaar-Based Payment Frauds Are Ballooning," *Morning Context*, May 30, 2022, https://themorningcontext.com/internet/aadhaar-based-payment-frauds-are-ballooning.

31. Vishal Krishna, "This Startup Clocked $9M in Revenue by Lending to Just Students and the Self-Employed," *YourStory*, March 30, 2021, https://yourstory.com/2021/03/kolkata-based-lending-platform-mpokket-students-instant-loan.

32. Srikanth Laxmanan, interview by Rahul Mukherjee, October 31, 2021.

33. Gopal Sathe, "How Sai Baba Was Made to Spy on Your Phone for Credit Ratings," HuffPost, July 3, 2019, https://www.huffpost.com/archive/in/entry/fintech-apps-privacy-snooping-credit-vidya_in_5d1cbc34e4b082e55373370a.

34. Srikanth Lakshmanan, interview.

35. Snigdha Poonam and Samarth Bansal, "Inside India's Booming Dark Data Economy," *Rest of World*, December 20, 2020, https://restofworld.org/2020/all-the-data-fit-to-sell/.

36. Kailash Nath, "Why We Invested in CreditVidya," Medium, May 15, 2019, https://medium.com/bharat-innovations-fund/why-we-invested-in-creditvidya-18a3b404af40.

37. Langevin, "Big Data." Also see Privacy International's report "Fintech's Dirty Little Secret: Lenddo, Facebook, and the Challenge of Identity," October 23, 2018, https://privacyinternational.org/long-read/2323/fintechs-dirty-little-secret-lenddo-facebook-and-challenge-identity.

38. Sahib Sharma and Vishwanath Nair, "Lenddo to Help Indian Lenders Assess Small Value Customers Better," *Mint*, July 20, 2016, https://www.livemint.com/Industry/XTvRG717e6bq5XSz7lRVLJ/Lenddo-to-help-Indian-lenders-assess-small-value-customers-b.html.

39. Global NewsWire, "Lenddo Expands Credit Risk and Verification Technology Globally," January 21, 2015, https://www.globenewswire.com/news-release/2015/01/21/698849/34586/en/Lenddo. Jum Balea writes that "Lenddo uses algorithms that pull and analyze data from these people's [individuals from the emerging middle class] social media accounts—who their friends are, how often they interact, their interests—and turns the information into a rating or score that says if they're likely to pay back or default on their loans. Or in other words, if they are to be trusted." Jum Balea, "Lenddo Stops Lending, Now Helps Clients Determine Customer Trustworthiness," *TechInAsia*, January 25, 2015, https://www.techinasia.com/lenddo-customer-trustworthiness.

40. Langevin, "Big Data," 800.

41. Langevin, "Big Data," 802.

42. Langevin, "Big data," 799.

43. Alexandra Rizzi, Alexandra Kessler, and Jacobo Menajovsky, "The Stories Algorithms Tell: Bias and Financial Inclusion at the Data Margins," *Center for Financial Inclusion*, March 31, 2021, https://www.centerforfinancialinclusion.org/the-stories-algorithms-tell-bias-and-financial-inclusion-at-the-data-margins.

44. Arti Singh, "Inside the Scramble to Cut Off Chinese Loan Apps," *Morning Context*, February 3 2021, https://themorningcontext.com/internet/inside-the-scramble-to-cut-off-chinese-loan-apps.

45. Operation Hafta Vasooli on Zee Business channel (April 2020): https://www.youtube.com/watch?v=NMQdKi_zVJQ.

46. Pravin Kalaiselvan, interview by Rahul Mukherjee, September 8, 2022, Mumbai. It is crucial to note here that fintech companies developing questionable third-party data gathering tools and engaging in manipulative practices of digital lending do not just exist in China and Singapore, but that they are spread across several countries of South Asia, South East Asia, and East Africa.

47. Singh, "Inside the Scramble."

48. Some established and legitimate lending platforms disburse money raised by their own in-house NBFC arm, which is funded by promoter equity. For example, the loan apps Navi's and KreditBee's NBFC arms are Navi Finserv and KrazyBee, respectively. Navi and KreditBee do not always depend on Navi Finserv and KrazyBee but enter into colending arrangements with other NBFCs and banks. Gaurav Noronha, "Sachin Bansal's Flipkart Baggage Weighs Navi Down," *Ken*, June 19, 2023, https://the-ken.com/story/https-the-ken-com-story-sachin-bansals-flipkart-baggage-weighs-navi-down/.

49. Dieter et al., "Multi-Situated App Studies"; Carolin Gerlitz, Anne Helmond, David B. Nieborg, and Fernando N. van der Vlist, "Apps and Infrastructures: A Research Agenda," *Computational Culture: A Journal of Software Studies* 7 (2019), http://computationalculture.net/apps-and-infrastructures-a-research-agenda/.

50. See the HasGeeks presentation by Suman Kar and Srikanth Lakshmanan, "KillerApps: Detecting Predatory Fintech Apps," https://www.youtube.com/watch?v=jt08arX9EbE. In the presentation, Kar mentions that the source code APIs suggest that these microloan apps ask for numerous kinds of permissions, including access to flashlight, the phone's network state, and images and media on the phone and that they even ask to kill background processors.

51. Ramanathan, "The Unregulated Tech Supermarket."

52. Ramanathan, "The Unregulated Tech Supermarket."

53. Sophy Yang, "China Cracks Down on Malicious Lending and Web Crawlers, Temasek-Backed Tongdun Tech Implicated," *China Morning Post*, October 23, 2019, https://www.chinamoneynetwork.com/2019/10/23/china-crack-down-on-malicious-lending-and-web-crawlers-temasek-backed-tongdun-tech-implicated.

54. In 2021, through a single iOS software update, Apple curtailed third-party data gathering practices by restricting the ability of (non-Apple) apps to track users by offering an opt-out option for users from such forms of data collection. These privacy changes made by Apple obstructed the ability of rivals Meta and Google to target ads at consumers while Apple's own advertising business grew leaps and bounds. See Patrick McGee, "Apple's Privacy Changes Create Windfall for Its Own Advertising Business," *Financial Times*, October 17, 2021, https://www.ft.com/content/074b881f-a931-4986-888e-2ac53e286b9d.

55. Jennifer Pybus and Mark Coté, "Did you give permission? Datafication in the Mobile Ecosystem," *Information, Communication & Society* 25, no. 11 (2022): 1650–1668.

56. David Medine and Gayatri Murthy, "Making Data Work for the Poor: New Approaches to Data Protection and Privacy," *CGAP: Focus Note*. January 2020, https://www.cgap.org/sites/default/files/publications/2020_01_Focus_Note_Making_Data_Work_for_Poor_0.pdf.

57. Carolin Gerlitz, Anne Helmond, David B. Nieborg, and Fernando N. van der Vlist, "Apps and Infrastructures: A Research Agenda," *Computational Culture: A Journal of Software Studies* 7 (2019), http://computationalculture.net/apps-and-infrastructures-a-research-agenda/.

58. Jeremy Wade Morris, Robert Prey, and David Nieborg, "Engineering Culture: Logics of Optimization in Music, Games, and Apps," *Review of Communication* 21, no. 2 (2021): 161–175.

59. Kar and Lakshmanan presentation.

60. Nishant Kauntia, "How Predatory Loan Apps Run Fake Ads on Google, Facebook to Target Vulnerable Indians," MediaNama, September 28, 2021, https://www.medianama.com/2021/09/223-loan-apps-fake-ads-google-facebook/.

61. Safiya Umoja Noble notes, "Google Search is in fact an advertising platform," and further, succinctly adds, "Google creates advertising algorithms, not information

algorithms." Safiya Umoja Noble, *Algorithms of Oppression: How Search Engines Reinforce Racism* (New York: NYU Press, 2018), 38. The scammers have further shifted tactics by disbursing loans through APK files and WhatsApp links which do not necessarily lead to Google or Apple online stores. It is possible that a fraudulent loan app ad on Instagram can take borrowers to a web page which looks like Google PlayStore but is not. On a loan app's online page, they might see a photoshopped version of an NBFC license that certifies the loan app but which the NBFC is unaware of. See Soumyarendra Barik, "Fraud Loan Apps have a free run on social media platforms. Here's how," *The Indian Express*, December 27 2023. https://indianexpress.com/article/business/banking-and-finance/fraud-loan-apps-have-a-free-run-on-social-media-platforms-heres-how-9033678/

62. An almost unwritten rule is that the fintech loan app company can offer to cover up to 5 percent of the losses in a first loan default guarantee (FLDG) clause as part of the contractual agreement between the lending platform and the NBFC. During the pandemic, as investigative financial journalist Arti Singh wrote, some loan companies offered NBFCs "100% FLDG," which meant that the "fintech partner, which finds the actual borrowers, will take on all of the potential losses." See Arti Singh, "Inside the Scramble."

63. RBI Circular, "Guidelines on Digital Lending," Reserve Bank of India, September 2, 2022, https://rbidocs.rbi.org.in/rdocs/notification/PDFs/guidelinesdigitallendingd5c35a71d8124a0e92aeb940a7d25bb3.pdf.

64. Ashwin Manikandan, "Instant Digital Loans: Convenience Comes with a Heavy Cost," Morning Context, August 29, 2023, https://themorningcontext.com/internet/instant-digital-loans-convenience-comes-with-a-heavy-cost.

65. Tressie McMillan Cottom, "Where Platform Capitalism and Racial Capitalism Meet: The Sociology of Race and Racism in the Digital Society," *Sociology of Race and Ethnicity* 6, no. 4 (2020): 441–449. "Predatory inclusion is the logic, organization, and technique of including marginalized consumer-citizens into ostensibly democratizing mobility schemes on extractive terms."

66. I am indebted to Marc Steinberg and Joshua Neves for the way they formulate "convenience" to include aspects of logistics, service, and affect that need the tools of both political economy and cultural studies for holistic comprehension. See Marc Steinberg and Joshua Neves, eds. introduction to *In/Convenience: Inhabiting the Logistical Surround* (Amsterdam: Institute of Network Cultures, 2023), 5–10.

67. Amit Das, interview by Rahul Mukherjee, July 3, 2024.

68. Beni Chugh, Varun Agarwal, and Malvika Raghavan, "Indians Care Deeply About Privacy of Personal Data, Shows New Study," *Business Standard*, December 15, 2017, https://www.business-standard.com/article/current-affairs/indians-care-deeply-about-privacy-of-personal-data-shows-new-study-117121500346_1.html.

69. Beni Chugh, interview by Rahul Mukherjee, September 2022, Mumbai.

70. Karim, *Microfinance and Its Discontents*; Langevin, "Big Data."

71. Jacqueline Hicks, "Digital ID Capitalism: How Emerging Economies Are Re-inventing Digital Capitalism," *Contemporary Politics* 26, no. 3 (2020): 330–350.

72. Mugdha Variyar, "Rang De Helps Rural Women Use India Stack to Avail Micro Loans," *Economic Times*, March 7, 2017, https://economictimes.indiatimes.com/small-biz/startups/rang-de-helps-rural-women-use-india-stack-to-avail-micro-loans/articleshow/57520901.cms?from=mdr.

73. Devashish Mulye, "What Is OCEN and What It Means for Digital Lending," *FinBox*, July 27, 2017, https://finbox.in/blog/what-is-ocen-embedded-credit-digital-lending/.

74. Chris Clarke writes, "When providers of credit are enabled to constitute a form of 'infrastructure' in this way, this allows for particular groups and businesses, in this case the platform lending industry, to profit at the expense of people who become increasingly indebted as a result of the discourses and practices surrounding financial inclusion policies." Chris Clarke, "Platform Lending and the Politics of Financial Infrastructures," *Review of International Political Economy* 26, no. 5 (2019): 863–885.

Conclusion

1. Vanita Khandekar-Kohli, *The Indian Media Business: Pandemic and After* (New Delhi: Sage, 2021), xx.

2. Anand Venkatanarayanan, "WhatsApp Missed Call Job Scam: Why It's Happening and How Can It Be Stopped," *Indian Express*, May 19, 2023.

3. The discussion of the entanglement of transportation and communication technologies in the mobilities literature in this book would have benefited from examples of food-delivery workers, who are often migrants from rural areas to urban locations in India. Unfortunately, this is beyond the scope of this book. These workers depend on mobile phone apps (as much as their customers who order food) and bikes/cars/cycles to deliver food on time. The opacity of phone app algorithms in allocating delivery assignments tends to exploit the already informal nature of delivery work in the Global South, but workers have found ways of gaming the algorithm. See Rida Qadri and Noopur Raval, "Mutual Aid Stations," *Logic Magazine* 13 (2021), https://logicmag.io/distribution/mutual-aid-stations/.

4. On the contradictions and double binds marking the implementation and experiences of the digital, refer to Mohamed Zayani and Joe F. Khalil, *The Digital Double Bind: Change and Stasis in the Middle East* (Oxford: Oxford University Press, 2024).

5. Serena Natile, *The Exclusionary Politics of Digital Financial Inclusion: Mobile Money, Digital Walls* (London: Routledge, 2020), 11.

6. Refer to Ayush Wadhwa, interview with Yashoraj Tyagi, "The Reality of Indian Fintech? Ft. Yashoraj Tyagi CEO of Cashe," April 27, 2024, https://www.youtube

.com/watch?v=gNLAJW2_7Ik. Also see Akshay Mehrotra's speech in IBS Intelligence, "Reimaging the Credit Ecosystem," panel discussion, Payments & Lending Conference, March 24, 2022, https://www.youtube.com/watch?v=oW3S8J58LWo.

7. Jenna Burrell and Marion Fourcade, "The Society of Algorithms," *Annual Review of Sociology* 47 (2021): 223.

8. Purnima Mankekar, *Unsettling India: Affect, Temporality, Transnationality* (Durham, NC: Duke University Press, 2015), 190.

9. Rune Bolding Bennike, Mattias Borg Rasmussen, and Kenneth Bo Nielsen, "Agrarian Crossroads: Rural Aspirations and Capitalist Transformation," *Canadian Journal of Development Studies* 41, no. 1 (2020): 40–56.

10. Uncertain Commons, *Speculate This!* (Durham, NC: Duke University Press, 2013).

11. Kavita Philip, "Telling Histories of the Future: The Imaginaries of Indian Technoscience," *Identities* 23, no. 3 (2016): 276–293.

12. Tim Unwin, "Digital Economies at Global Margins: A Warning from the Dark Side," in *Digital Economies at Global Margins*, ed. Mark Graham (Boston: MIT Press, 2019).

13. Laura Bear, Ritu Birla, and Stine Simonsen Puri, "Speculation: Futures and Capitalism in India," *Comparative Studies of South Asia, Africa and the Middle East* 35, no. 3 (2015): 387–391.

14. Ananya Roy, "Subjects of Risk: Technologies of Gender in the Making of Millennial Modernity," *Public Culture* 24, no. 1 (2012): 131–155.

15. Naomi Fry, "The Comforts of 'WeCrashed' and the Modern Grifter Series," *New Yorker*, April 4, 2022, www.newyorker.com/magazine/2022/04/04/the-comforts-of-wecrashed-and-the-modern-grifter-series. Lisa Parks and France Winddance Twine, "Millennial Messiahs, Female Fixers, and Corporate Boards: Workplace Power Dynamics in Tech TV Dramas," *Film Quarterly* 76, no. 3 (2023): 25–35.

16. Samarth Bansal notes this during an interview in the documentary *India's Thriving Scam Industry: Before You Call Tech Support* (2022), https://www.youtube.com/watch?v=7CZReZ24-to.

17. Deepu Sebastian Edmond, interview by Rahul Mukherjee, June 10, 2025.

18. Harish Damodaran, "The Unaspiring Millions," *The Indian Express*, October 26, 2024. https://indianexpress.com/article/opinion/columns/harish-damodaran-writes-why-aspiration-is-dead-in-india-9639120/.

19. Deepu Sebastian Edmond, interview by Rahul Mukherjee, June 10, 2025.

20. NDTV, "Caste Set Aside by Aspiration," June 4, 2010, https://www.youtube.com/watch?v=mPr7wTK6xZM

21. Ajeet Kumar Pankaj and Manoj K. Jha, "The Making of New Dalit Middle Class: Struggle, Aspirations, and Changing Lifestyle in Navi Mumbai, India," *Journal of Asian and African Studies* 59 (8): 2543–2555.

22. Anant Kamath, "'Untouchable' Cellphones? Old Caste Exclusions and New Digital Divides in Peri-urban Bangalore," *Critical Asian Studies* 50, no. 3 (2018): 375–394.

23. Shadma Shaikh, interview by Rahul Mukherjee, August 4, 2019. This unfolds amid a landscape where some Indian tech geeks, consumer rights organizations, and digital security experts like Nikhil Pahwa (Media Nama) and Kiran Jonnalagadda (HasGeek) continue to protest against Aadhaar leaks and so on.

24. Beni Chugh, Varun Agarwal, and Malvika Raghavan, "Indians Care Deeply About Privacy of Personal Data, Shows New Study," *Business Standard*, December 15, 2017, https://www.business-standard.com/article/current-affairs/indians-care-deeply-about-privacy-of-personal-data-shows-new-study-117121500346_1.html.

25. Payal Arora, *The Next Billion Users: Digital Life Beyond the West* (Cambridge, MA: Harvard University Press, 2019), 180.

26. Arora, *The Next Billion Users*, 155

27. Arora, *The Next Billion Users*, 175–176.

28. Srirupa Roy, "Being the Change: The Aam Aadmi Party and the Politics of the Extraordinary in Indian Democracy," *Economic and Political Weekly* 49, no. 15 (2014): 49.

29. Press Trust of India, "Aam Aadmi Party's Sting App to Help Fight Corruption," NDTV, April 3, 2015, https://www.ndtv.com/delhi-news/aam-aadmi-partys-sting-app-to-help-fight-corruption-751971.

30. Ravi Sundaram, "Publicity, Transparency, and the Circulation Engine: The Media Sting in India," *Current Anthropology* 56, no. 12 (2015): S297–S305.

31. Sundaram, "Publicity, Transparency, and the Circulation Engine," S297. The "unedited" entrapment videos were low-resolution JPEG and AVI files, which according to Ravi Sundaram had an aesthetic of "barely visible individuals," "less than audible sound," and "shoddy camera-work," but which because of such an aesthetics, or despite it, were able to invite "public forensics."

32. Furthermore, a particularly potent version of Pegasus seems enabled with "zero click" exploit ("requires no interaction with the target and leaves no visible evidence of tampering") that does not need the targeted victim/user to click on a malicious link/text message/email to become activated. After surreptitiously taking over the control of the surveilled user's phone, Pegasus spyware leaves the device's vendor/operating system (such as Apple/iOS) totally unaware: This phenomenon is referred to as "zero day" vulnerability. For more, refer to Ron Deibert, "Protecting Society from Surveillance Spyware," *Perspectives*, Winter 2022: 15–17.

33. Akhil Gupta, "The Future in Ruins: Thoughts on the Temporality of Infrastructures" In Nikhil Anand, Akhil Gupta, and Hannah Appel (eds.), *The Promise of Infrastructure* (Durham, NC: Duke University Press, 2018).

34. Clarke, "Platform Lending and the Politics of Financial Infrastructures," 866.

35. Pramod Varma, "Building for a Billion: Aadhaar & India Stack Experience," *ConfEngine* November 21, 2021, https://www.youtube.com/watch?v=9oymcn5PTdA.

36. In a country where the homeless, construction workers, and migrant laborers toil under extreme conditions, hands wear out, thus foreclosing the possibility of stable fingerprints.

37. Amber Sinha and Shruti Trikanad, "Linking Caste and Surveillance in India: How Digital Governance Has Legitimised Caste Discrimination," in *The Cambridge Handbook of Race and Surveillance*, ed. Michael Kwet (Cambridge, UK: Cambridge University Press, 2023), 76–94.

38. "When Aadhaar is seeded into every database, it becomes a bridge across discreet data silos, which allows anyone with access to this information to reconstruct a profile of an individual's life." See Supreme Court of India, *Justice K.S. Puttaswamy (retd) and Another* v *Union of India and Others*, WP (C) 494/2012 (2018). September 26, 2018, https://main.sci.gov.in/supremecourt/2012/35071/35071_2012_Judgement_26-Sep-2018.pdf. A person's Aadhaar number would also need to be linked to various rehabilitation and welfare schemes if that person wants to receive governmental subsidies, reimbursements, benefits, and relief, which over time would allow for a reconstruction of the person's profile. Gautam Bhatia, "The Aadhaar Judgment: A Dissent for the Ages," Indian Constitutional Law and Philosophy, September 27, 2018, https://indconlawphil.wordpress.com/2018/09/27/the-aadhaar-judgment-a-dissent-for-the-ages/. Ashish Rajadhyaksha, "The Ambivalence of 'Creep': A Biopolitical History of Digital Times," in *Overload, Creep, Excess—An Internet from India*, ed. Nishant Shah, Ashish Rajyadhaksha, and Nafis Aziz Hasan (Amsterdam: Institute of Network Cultures), 88.

39. Ursula Rao and Vijayanka Nair, "Aadhaar: Governing with Biometrics," *South Asia: Journal of South Asian Studies* 42, no. 3 (2019): 469–481.

40. Jonathan Ananda, "1.5 Crore Unregistered Construction Workers in the Lurch sans Relief," *New Indian Express*, March 27, 2020, https://www.newindianexpress.com/business/2020/mar/27/15-crore-unregistered-construction-workers-in-the-lurch-sans-relief-2122056.html#. There have been reports of starvation deaths because Aadhaar data could not authenticate the fingerprints of legitimate beneficiaries who were thus deprived of timely rations that they were legally entitled to receive. See Aarefa Johari, "Yet Another Aadhaar-Linked Death? Denied Rations for 4 Months, Jharkhand Woman Dies of Hunger." *Scroll,* February 3, 2018, https://scroll.in/article/867352/yet-another-%20aadhaar%20linked-death-jharkhand-woman-dies-of-hunger-after-denial-of-rations.

41. Anja Kovacs and Nayantara Ranganathan, "Data Sovereignty, of Whom? Limits and Suitability of Sovereignty Frameworks for Data in India," *Internet Democracy Project Report*, November 2020, https://internetdemocracy.in/policy/policy-brief-data-sovereignty-of-whom-limits-and-suitability-of-sovereignty-frameworks-for-data-in-india.

42. Nayantara Ranganathan, "The Economy (and Regulatory Practice) That Biometrics Inspires: A Study of the Aadhaar Project," in *Regulating Biometrics: Global Approaches and Urgent Questions*, ed. Amba Kak (New York: AI Now Institute, 2020), 53.

43. Revati Prasad, "People as Data, Data as Oil: The Digital Sovereignty of the Indian State," *Information, Communication & Society* 25, no. 6 (2022): 808. Some scholars have understood this to be biometric capitalism which has taken different forms in different countries such as India, South Africa, Ghana, and Kenya. Biometric financialization has been implemented in South Africa much earlier than India through smart cards (run by Net1) which utilized fingerprint biometrics for identification and verification and disbursed social grants, pensions, and various credit products. Fingerprinting has a long history in South Africa across colonial and Apartheid periods. So, phones or smartphones are not necessarily required for biometric cash transfers or mobile credit infrastructures as the South African case suggests, which found takers such as the Ghanian Central Bank who devised a plan for biometrically controlled money supply (e-Zwich) through Net1 smart card technologies. Indeed, the smart cards are miniature programmable computers that are easy to carry, hence portable/mobile, and their functionalities can be replicated on phones. See Keith Breckenridge, "Biometric Capitalism: Infrastructures of Identification and Credit Risk on the African Continent in the 21st Century," *Technosphere x Knowledge*, April 2016. https://mediathek.hkw.de/en/video/truth-measures—biometrischer-kapitalismus—infrastrukturen-der-identifizierung-und-kreditrisiko-auf-dem-afrikanischen-kontinent-im-21—jahrhundert-von-keith-breckenridge?overlayObjectType=video&overlayObjectSlug=50469&type=video&slug=50469&backscreen=

44. From the use of India Stack and Aadhaar-enabled payment systems in governing populations to deployment of CCTV cameras and face recognition technologies in policing operations, the Indian state has worked extensively with start-up companies as part of "public private partnerships" (PPPs). These PPPs have created a situation where, as Shivangi Narayan writes, neither the government nor the private sector are ready to take responsibility for the consequences of these surveillance technologies on affected people. The tech entrepreneurs and engineers who are pushing surveillance tech to the government often eschew any exclusionary or discriminatory effects of such technologies by attributing the idea of neutrality and efficiency to their developed technological products. See Shivangi Narayan, "'For Your Own Safety': The Soft Push of Surveillance by the Private Sector in India," in *Policing and Intelligence in the Global Big Data Era*, Vol. I: *New Global Perspectives on Algorithmic Governance*, ed. Tereza Østbø Kuldova Helene, Oppen Ingebrigtsen Gundhus, and Christin Thea Wathne (Switzerland: Palgrave McMillan, 2024), 181–212.

45. Prasad, "People as Data."

46. Nick Couldry and Ulises A. Mejias, *The Costs of Connection: How Data Is Colonizing Human Life and Appropriating It for Capitalism* (Stanford, CA: Stanford University Press: 2019), 188.

47. Catriona Gray, "More than Data Extraction: Rethinking Data's Colonial Political Economy," *International Political Sociology* 17, no. 2 (2023): 1–20.

48. Abhijit Das, "India Faces Threat of Digital Colonization," *Hindu Business Line*, January 12, 2020, https://www.thehindubusinessline.com/opinion/india-faces-threat-of-digital-colonisation/article30551143.ece.

49. Data Protection Bill: https://prsindia.org/files/bills_acts/bills_parliament/2019/Personal%20Data%20Protection%20Bill,%202019.pdf. While the recent data protection law tabled for discussion in the parliament eased some restrictions/compulsions about data localization within India, the final shape of the bill remains to be seen.

50. Other than Mumbai and Chennai, data centers have also been constructed in Delhi (and the Noida region), Hyderabad, and Bengaluru. See Todd Olsen, "What's Next for Data Centers," *Cushman and Wakefield*, August 7, 2021, https://www.cushmanwakefield.com/en/india/insights/whats-next-for-data-centres-in-india.

51. India's main stock exchange is located in Mumbai—hence, the need for lightning-fast financial data transmissions.

52. See CRISIL report: "Demand for Data Centres Goes Viral," https://www.communicationstoday.co.in/demand-for-data-centres-goes-viral-crisil-research.

53. Salil Panchal, "This Real Estate Asset Is Pandemic Proof," *Forbes India*, June 29, 2021, https://www.forbesindia.com/article/real-estate-special/this-real-estate-asset-is- pandemic-proof/68819/1.

54. Daniel Greene, "Landlords of the Internet: Big Data and Big Real Estate," *Social Studies of Science* 52, no. 6 (2022): 904–927. Even as streaming services like Netflix and SonyLiv and social media platforms like Google and Facebook have their own dedicated data centers, they also rent our space from internet landlords like Equinix and Digital Realty.

55. While some of these data center projects might exist as islands, they are proximate to transport corridors, a newly constructed airport, and the expanding port. They also receive a government-supported uninterrupted power supply at industrial rates from fossil fuels and hydropower. Navi Mumbai is seismologically safer than Mumbai, and unlike Mumbai, which is prone to waterlogging and flooding, Navi Mumbai has storm-water management and better catchment areas. See Manish Jha and Rishi Jha, "Transforming Urban Form and Emergent Statecraft: Data Centers in Navi Mumbai," in *Data Centers as Infrastructures: Frontiers of Digital Governance in Contemporary India*, ed. Manish Jha and Ritam Sengupta (Hyderabad, India: Orient Blackswan, 2022), 144.

56. Javiera Barandiarán, Mona Damluji, Stephan Miescher, et al., "Energy Justice in Global Perspective: An Introduction," *Media+Environment* 4, no. 1 (2022), https://doi.org/10.1525/001c.37073.

57. This is not the only way of extracting lithium. In some places, open-pit mining is also involved.

58. Brian Pereira, "NTT Ltd. Launches 7th Data Center in Mumbai and Solar Plant in Maharashtra," *Digital Creed*, October 1, 2020, https://www.digitalcreed.in/ntt-data-center/.

59. David Singh, "'This Is All Waste': Emptying, Cleaning and Clearing Land for Renewable Energy Dispossession in Borderland India," *Contemporary South Asia* 30, no. 3 (2022): 402–419.

60. Ned Rossiter and Soenke Zehle, "Logistical Media and Environmental Media: Energy, War, and Politics of Distribution," *Journal of Environmental Media* 4, no. 1 (2023): 78. Beyond material limits of land and energy encountered by mobile streaming/data centers, there are entangled ethicopolitical concerns. The philosopher of technology Langdon Winner wrote a book with the subtitle "A Search for Limits in an Age of High Technology" to ask the question that amidst the "inexhaustible power of scientific technology" where will we "draw the line." He notes that while people recognize the need for limits on certain technologies based on risks to public health or environmental degradation or natural resource crunches, these criteria do not sufficiently account for the possible political and moral limits for technological change. See Langdon *Winner, The Whale and the Reactor: A Search for Limits in an Age of High Technology*, Chicago and London: University of Chicago Press, 1986, p. xi. I am not suggesting that countries in the Global South, especially those who have hot and wet climates should not have data centers. Rather, environmental issues around data centers need to be carefully considered, and one should question temperature standards of data center air-conditioning which have a Global North (Western) bias, and think relationally about thermal autonomy and data sovereignty. For more on this see Nicole Starosielski, *Media Hot and Cold* (Durham: Duke University Press, 2022).

61. Pranav Malhotra and Rich Ling, "Agency Within Contextual Constraints: Mobile Phone Use Among Female Live-Out Domestic Workers in Delhi," *Information Technologies & International Development*, 16: 32–46. Also see Arora, *The Next Billion Users.*

62. Meghna Gupta, Devansh Mehta, Anandita Punj, Indrani Medhi Thies, "Sophistication with Limitation: Understanding Smartphone Usage by Emergent Users in India," COMPASS '22, June 29–July 1, 2022, Seattle, WA, https://www.microsoft.com/en-us/research/uploads/prod/2022/05/compass22-34-taps.pdf.

63. Shrinath V., "Making Opportunities Inclusive for First Time Digital Users," in *Feminist Futures of Work: Reimagining Labor in Digital Economy*, ed. Payal Arora et al. (Amsterdam: Amsterdam University Press, 2023).

64. Sirpa Tenhunen, *A Village Goes Mobile: Telephony, Mediation, and Social Change in Rural India* (Oxford: Oxford University Press, 2018). Anja Kovacs, "'Chupke, Chupke': Going Behind Mobile Phone Bans in Rural India," Gendering Surveillance: Internet Democracy Project, February 2017, https://genderingsurveillance.internetdemocracy.in/phone_ban/.

65. Tereza Capelos and Ipshita Basu, "Who Is in the Middle: Social Class, Core Values, and Identities in India," *Political Psychology* 43, no. 1 (2022): 89. Rahul Gandhi and the Congress party were also interested in this group as an important vote bank in the runup to the 2014 elections, though they did not use the "neo-middle" phrasing. They discussed this group as a "sandwiched class" between the rich and poor and emerged with the acronym NRMB: "not rich, not middle-class, not below poverty line." See Rajeev Deshpande, "Congress Poll Strategy: Rahul Gandhi Eyes 'Sandwiched Class,'" *Times of India*, January 24, 2014, https://timesofindia.indiatimes.com/india/congress-poll-strategy-rahul-gandhi-eyes-sandwiched-class/articleshow/29271170.cms.

66. Deshpande, "Congress Poll Strategy," 91. While income levels and professions might separate traditional middle classes and neo-middle aspirers, middle-class identity is also about "acquired performative experience embedded in a social setting, shaped through intersubjective relations with the other classes."

67. While there are significant overlaps among the "neo–middle classes" and "neomobiles," bonded as they are by aspirational politics in India, it would be tautological to suggest a neat formulaic equivalence of these two groups. Welfare beneficiaries are a key part of the neomobiles because they receive several benefits in digital accounts on their newly acquired phones. However, many neo–middle-class members do not meet the criteria for receiving benefits because they are not below the poverty line. Rukmini S. mentions two striking data points that should caution us towards drawing an easy correlation between mobile phone ownership and being middle class. From 2004-5 to 2011-12, cell phone ownership propelled from 7% of the population to 82%, but only 3% population owned all five basic consumer goods: motor vehicles, TV, refrigerator, cooler/AC, and laptop/computer. See Rukmini S., *Whole Numbers and Half Truths: What Data Can and Cannot Tell Us About Modern India* (Chennai: Context, 2021), 143.

68. Rashmi M., "Offline Media Economy: Digitally Marginalized Users of Mobile Phones," in *The Indian Media Economy*, vol. II, *Market Dynamics and Social Transactions*, ed. Adrian Athique and Vibodh Parthasarathi (Oxford: Oxford University Press, 2018), 208–209.

69. Hari Krishna, "Can Stage Become the Netflix for Bharat," *The Ken*, April 25, 2025, https://the-ken.com/newsletters/two-by-two/can-stage-become-the-netflix-for-bharat/

70. Swartz, *New Money*.

71. Momin, "Lumpen Utopia," 221.

Index

Note: Page numbers followed by an *f* denote a figure.

4G LTE technology, 84

Aadhaar database, 30
  critics of, 228
  data, mobility of, 229
  Digital India and, 36–37
  digital lending and, 196
  fraud and, 198
  government promotion of, 155
  interlinked digital platforms, 191
  marginalized communities and, 190
  phishing and, 169
  selfies and, 32
  surveillance and, 226–228, 229, 290n44
Aakanksha, 27
Aam Aadmi Party, 225
Abundance, 3, 5, 6, 242n10, 252n116
Access doctrine, 12
Active audiences, 122
Actor-network theory (ANT), 57–58
Addressability, 112
Advance AI, 207
Advani, Prakash, 106–108, 107*f*
Advertising
  digital lending and, 187, 188*f*, 189*f*
  influencer marketing, 151–152
  microloan apps, 189, 209, 210*f*–211*f*
  mobile phones and, 1
  political, 117–118
  TikTok and, 132–134, 133*f*–134*f*, 150
  unlimited data plans, 33–34, 34*f*
  virality and, 2, 132–134, 150
Advertising-based video-on-demand (AVoD), 86, 87, 239
Affect, 15, 78, 266n21
  approaching/crossing limits and, 8
  anticipation as a state of, 8
  call center vocalizations and, 163
  circulation of, 122–124, 149
  habits and, 129, 268n43
Airtel, 2, 31, 174*f*
  bundling and, 83, 218
  content partnerships, 97–98
  as distributor, 98–100
  unlimited data plans, 33–34, 34*f*
  Xstream feature, 83, 98, 99*f*
Akamai Technologies, 89, 91, 101, 102, 103–106
Akhlaq, Mohammad, 127–128
Ali, Mohammad, 128
ALTBalaji, 84, 87, 95, 97, 112, 116
Alternative finance. *See* Microloan apps, predatory
Amazon CloudFront, 87, 91, 103, 106
Amazon Prime Video, 84, 100, 112
Amazon Web Services (AWS), 101

Ambedkar, Bhimrao, 136
Amnesty International, 226
Anticipation, 8–9, 115, 227
Anticorruption, surveillance and, 225
Appadurai, Arjun, 22, 130
Application programming interfaces (APIs), 30, 32, 207*f*, 208
App stores. *See* Google Play store
Arceneaux, Noah, 17
Archambault, Julie Soleil, 24
Archiving practices, 63–64
Arora, Payal, 22–23, 225
Arthur, Tori Omega, 114
Artificial intelligence (AI), 138, 152, 207, 252n116, 273n102
Aspirational politics, 21–28, 249n79
  aspirational hatred and, 130
  cybercrime and, 159
  data and energy, 236
  limits of, 219
  mobile media distribution and, 16, 28–33, 217
  viral advertising and, 150
Aspirational realness, 148
Aspirations
  ambivalence about, 221
  caste system and, 222–223
  of content creators, 6–7, 28, 147–148, 150–151
  cybercrime and, 175–178
  digitization, expansion of, 164
  entrepreneurship, 8, 25, 27, 131
  gendered, 179
  loans and, 195, 214, 281n20
  mobility, limitations of, 68–71, 221
  narratives, 220–224
  neoliberal politics and, 25–26
  neo-middle class and, 21–22, 35, 217, 220–221, 228, 238
  in Skill India Digital Platform, 8–9
  unrealistic, 6–7
Athique, Adrian, 37
Audiences, 91–96, 217–218, 239
  analysis of, 138–139
  new-to-mobile-phone, 95
  segmentation of, 139
  streaming media and, 86
  virality, effect on, 121–122
Audio players, 71, 72*f*–75*f*, 76, 77*f*
Autopoiesis, 129

Bachchan, Amitabh, 1
Badshah, 132, 148
Baishya, Anirban, 124
Balaji Telefilms, 97
Bansal, Samarth, 200, 222
Barad, Karen, 58
Barandiarán, Javiera, 235
Barnett, Kyle, 142
Barney, Darin, 16
Barry, Ellen, 22
Beaster-Jones, Jayson, 55, 56
Bedi, Seher, 145–146, 147
Bedi, Umang, 119
Berlant, Lauren, 7
BHIM (Bharat Interface for Money), 37
Bhojpuri music, 23, 45–46, 52–55, 78, 253n2
BigoLive, 36, 136
Big Tech, 30, 40, 87, 120, 190, 214–215, 230. *See also* Platforms
Bihari Dalit youth, 78–79
Biometric markers, 198, 228–229, 290n43
Black money, 156, 183–184
Blockbuster, 59–61
Bollywood celebrities, 146–147
Bollywood films, 131–132
Bouquillion, Philippe, 101–102
Brass, Paul, 129
Braun, Joshua, 29, 88, 149
Brock, André, 33
BTS, 35, 139
Buffering, 92–93, 113

Bundling, 83–84, 96, 97–100, 218
Burrell, Jenna, 220

Cable operators, local, 112, 115, 263–264n65
Cable-wallas, 112, 115
Call center industry, 158, 161–163
Callon, Michel, 57
Cambodia, 183
Capitalism, 5–6, 30, 150, 157–158, 214, 230–231, 235–236
Carriage fees, 97
CarryMinati, 135
CashBus India, 209
Cashless economy, 11
Caste system
  aspirational narratives and, 222–223
  audience segmentation and, 139
  Bhojpuri music and, 78
  class mobility, 222–223
  cybercrime and, 175–176, 179–180
  digital lending and, 220
  in Gaya, 76–78
  hatred, spreading on social media, 135–136, 150
  mobility, limitations of, 71–79
  socioeconomic mobility and, 238
Cellular infrastructure, 84–85, 92–94
Censorship, 52–53, 64–65
CGAP (Cumulative Group to Assist the Poor), 208
Chacko, Priya, 27, 130, 170
Chalaby, Jean K., 114
Chamberlain, K. C., 70
Chat apps, 181. *See also specific apps*
Chattopadhyay, Swati, 69
Chinese DVD players, 60, 61*f*, 71
Chingari (video platform), 145
Chinnadurai, Sathana, 23–24, 237
Chip loading, 64. *See also* Memory cards
choke points, 3
Christopher, Nilesh, 135–136, 237
Chugh, Beni, 213, 224
Chun, Wendy, 129
Circumvention technologies, 60, 62–63
Class mobility. *See* Socioeconomic mobility
Coates, Jamie, 152
Colonialism, 230–231
Colors, 90–91
Communal riots, 122, 129
Commuters, streaming media and, 92–94, 113
Conditional-access system (CAS), 112
Consumer rights organizations, 192, 195–196, 209, 211
Content, movement of, 20
Content creators. *See also* Platform labor
  adaptability, 113
  aspirations of, 6–7, 28, 147–148, 150–151
  Chinnadurai and, 23–24
  content delivery networks and, 114
  content distributors, interactions with, 89
  Dalit Bahujan ("lower-caste") women, 8
  intermediaries, 149
  localization of, 139, 151
  performance targets, 136–137
  platform economy and, 11
  recruitment of, 138–139, 142–146, 143*f*–144*f*, 145*f*, 218
  retreats and workshops for, 146–147
Content delivery networks (CDNs), 5
  barriers to entry, 103–104
  cache model, 108–109, 111
  content creators and, 114
  dominant companies, 103–104
  hyperlocal, 12, 85–86, 104
  infrastructure, 102–104
  last-mile operators, 112
  local, 12, 106–109, 107*f*, 110, 115
  local cable operators and, 18

Content delivery networks (CDNs) (cont.)
  local ISPs and, 104–112, 107*f*
  subscription video-on-demand (SVoD) and, 102–103
  transmission paths, optimization of, 111
  Voot and, 91
Content moderation, 57–58
Convenience, 98, 156–157, 161, 191, 261n34
Cooper, Timothy, 6
Corruption, 183–184
Couldry, Nick, 230
COVID-19 pandemic, 83, 112, 182–183, 203
Cow vigilante groups, 6, 125–131, 126*f*, 127*f*, 149
Creator-centric approaches, 149
Credit card fraud. *See* Cybercrime
CreditVidya, 199–201
Cresswell, Tim, 78
Cricket, 84, 100–101, 102, 103*f*
Cryptocurrency, 183
Cuba, 52, 79–80
Cultural homogenization, 87
Cultural imperialism, 87
Cybercafes, 12
Cybercrime. *See also* Phishing; Scams
  advance fee scams, 180–181
  aspirational narratives and, 174–180, 222
  aspirational politics and, 159
  capitalism and financial scams, 157–158
  caste and, 175–176, 179–180
  chat apps and, 181
  coercion and human trafficking, 182–183
  during COVID-19 pandemic, 182–183
  credit card fraud, 156, 157–158, 160–161, 167–168
  cyber-fraud, 159
  "cybers," 164, 166
  demonetization and, 180
  Digital India and, 160, 164–166
  emotional blackmail, 180
  financial inclusion and, 168–169
  gender and, 175–176, 179–180
  lottery scams, 158, 161, 180–181
  mobile money services and, 156–157
  mobile phones and, 166–167, 169, 172–173, 179
  regulatory limits and, 184–185
  rise of, 24–25
  rural areas, 158–159
  vishing, 160–161, 164, 166–167, 174
  work-at-home job scams, 181

Dalit community, 8, 12–13, 125, 135–136, 223. *See also* Bihari Dalit youth
Damodaran, Harish, 222
Dance videos, 132–134, 133*f*–134*f*
Das, Amit, 212
Data capitalism, 30
Data centers, 230–236, 232*f*, 233*f*, 234*f*
Data collection, 38. *See also* Surveillance
  biometrics and, 198, 228–229
  data centers and, 230–236
  microloan apps, 192, 193, 199–201, 214
  mobility and, 224
  public attitudes toward, 213
  regulation of, 229–230
  social loan quotient, 220
Data colonialism, 230–231, 235–236
Datafication, 20
Data Protection Bill, 231
Dawkins, Richard, 122
Debt, 41, 192, 202, 212, 281n20
Deceptive lending practices, 203–206
Demonetization drive, 11, 20, 124
  black money and, 184
  coercive encouragement, 169–170

cybercrime and, 180
financial inclusion and, 155–156, 190–191
fintech and, 190
history of, 195–198
mobile money services and, 156–157
surveillance and, 226
Desai, Karishma, 8, 9
Dieter, Michael, 194
Digital capitalism, 231
Digital divide, 15
Digital ID capitalism, 214
Digital inclusion, 15
Digital India, 5, 24, 26, 27, 33–38
aspirational politics and, 34, 36
cybercrime and, 160, 164–166
digital money and, 155–156
infrastructure of, 169
rural areas, 173
vision for, 36–37
Digital infrastructures, 4–5, 7–8
Digitalization, of music, 50–51
Digital lending, 195–200. *See also* Microloan apps, predatory
advertising and, 187, 188*f*, 189*f*
alternative data, 198–199
apps and infrastructures, relationship between, 206–207
caste and, 220
cultural form and affect, 212
customers, alternative data for assessing, 200–203
deceptive lending practices, 203–206
digital payments and, 42
financial inclusion and, 191
fintech infrastructures, 204, 206–212
fraud in, 187, 189*f*, 198
gender and, 202, 213–214, 220
government promotion of, 214–215
as a habit and cultural norm, 212
history of, 195–198
intimidation practices in, 203–206, 204*f*
introduction to, 187–195
poverty alleviation, 196–197
stricter guidelines for, 211–212
for unbanked people, 15, 21, 30, 41, 156, 168, 191, 193–194, 208, 218
Digital money, 193. *See also* Digital lending; Mobile money services
Digital payments, 20, 42, 195
demonetization and, 156
financial inclusion and, 191
gateway for, 196
in Global South, 20
Global South, 20
government promotion of, 37
India Stack, 197
mobile data distribution, 20–21
mobile phones and, 5
platform economy and, 11
scams and, 181
Digital sovereignty, 229
Digital transactions, 11
Disney+Hotstar, 100, 101–102, 112
Distribution infrastructures, 89, 110, 111, 115–116
Donoghue, Courtney, 29
Doordarshan, 115
Doron, Assa, 17
Download vendors, 18, 31, 46, 47*f*–48*f*, 53–59, 54*f*–55*f*
Duffy, Brooke, 148
Duniya, Shailesh Ki, 153
Dupahiya, 71
DVD players, 60, 61*f*, 71

The Economic Times, 133–134
Edge servers/computing, 102, 105–111, 107*f*, 115
Edmond, Deepu Sebastian, 163, 178, 222–223
Electronic know your customer (eKYC), 196, 198, 207, 214–215

End-to-end encryption, 125
Energy, data centers and, 234–236
English language, 137–138. *See also* Vernacular languages
Epoch.api, 208
Eros Now, 97
Experimentarium museum, 5–6

Facebook, 56, 57, 117, 137–138
Fair usage limits, 3
Fan communities, 122–123
Faroq, Gowhar, 130–131
Federici, Silvia, 41
Feminist relational ontologies, 58
Fibe advertising, 32
Financial inclusion
  aspirational narratives of, 30, 221
  cybercrime and, 168–169
  demonetization and, 155–156, 190–191
  digital lending and, 191
  governmental model of, 155–156
  government promotion of, 215–216, 227
  women and, 13–14
Findlay, Rosie, 148
Fintech, 190–191, 198, 201. *See also specific companies*
Fintech as a service (FaaS), 31–32, 194, 206–208, 207*f*, 215
Fintech Festival India, 31–32, 32*f*
FlashRupee, 191
Forbidden Stories, 226
Fourcade, Marion, 220
Freedom, 4, 7
"Future Human: What Is Your Limit?" (exhibit), 6

Ganti, Tejaswini, 113
Gatekeepers, 123–124
Gawer, Ananbelle, 56
Gaya. *See* Memory cards
Gender
  aspirations and, 179
  call centers and, 163
  cybercrime and, 175–176, 179–180
  digital lending and, 202, 213–214, 220
  financial inclusion and, 13–14
  mobile media and, 237–238
  social surveillance, 237–238
  socioeconomic mobility and, 238
  vernacular music videos and, 58
Geo-blocking, 3
Geographical mobility, 19, 21–22, 23, 70, 78–79
Ghosh, Bishnupriya, 15
Ghosh, Ishita, 31
Global capitalism, 150
Globalization, 109
Global North, 230–231
Global South
  data movement, 230–231
  intermediaries, 31
  mobile payments, 20
  mobile phones and, 12
  perspectives from, 224
  predatory microloan apps, 219
  surveillance and, 225
  telecom bundling, 83–84
Google, Android permission system, 199–200
Google CloudFlare, 91, 103
GooglePay, 156–157
Google Play store, predatory microloan apps, 189, 194–195, 198, 205, 209, 212
Government, 36–37, 155–156, 214–216, 227. *See also* Aadhaar database; Demonetization drive; Digital India
Govindrajan, Radhika, 129
Grameen Bank, 196–197, 213–214
Green capitalism, 235–236
Greene, Daniel, 12, 231
Gupta, Akhil, 227
Gupta, Neena, 187

Hafta Vasooli media campaign, 203–204, 204*f*
Hate, spreading on TikTok, 135–136, 150
Havens, Timothy, 121
Hicks, Jacqueline, 214
Hiranandani Group, 232
Hobis, Geoffrey, 80
Hollingworth, Steve, 197
Hotstar, 102
  Airtel and, 98–100
  bundling and, 83, 218
  content delivery networks and, 91
  launch of, 84
  legacy broadcasters and, 98
  Reliance Jio, relationship with, 100–102
  sports content, 100–102
  Star Network and, 90
Human trafficking, 183
Humphreys, Laura-Zoe, 79
Hutchison, 2

Idea Cellular, 2
Imitation publics, 33, 132, 136
The Indian Express, 163, 222
India Stack, 11–12, 30, 37, 191, 195–197, 214–215, 226–227
Inequality, technological solutions, 15–16. *See also* Financial inclusion
Influencer marketing, 151–152
Influencers, 6–7, 119, 121, 132–133. *See also* Content creators; Microcelebrities
Informality
  Digital intermediaries and, 29–31
  Memory cards and vernacular music, 46, 49
  Money scams and, 174
Information and communication technologies for development (ICT4D), 12
Information overabundance, 6
Instantaneous flare-up thesis, 129
InstaReel, 151, 153
Insurgency, 163, 277–278n49
Interclass trolling, 135
Intermediaries, 31–33, 56–57. *See also* Talent agents
  content creators, 149
  between content creators and platforms, 149
  in Global South, 31
  microloan apps and, 194–195, 216
  mobile money services and, 31–32
  platforms, 149, 220
  streaming media/platforms, 87–89
  Uber and YouTube as, 56
International Monetary Fund, 184, 227
Internet-distributed video services, 87. *See also* Streaming media/platforms
Internet service providers (ISPs), 85, 104–112, 107*f*, 115
Intimidation, as part of loan recovery 203–206, 204*f*
iROKOtv, 114
iSPIRT (Indian Software Products Industry Roundtable), 37, 196, 215
Ithurbide, Christine, 101–102

Jaffrelot, Christophe, 170
Jain, Simran, 6–7
Jamaica, 158, 161
Jamtara, cybercrime and, 157–160, 162–164, 165*f*–166*f*, 167–173, 171*f*, 174*f*, 178–179, 222
Jamtara: Sabka Number Aayega, 157, 162*f*
  aspirational narratives and, 159–161, 175–178, 177*f*, 179–180, 222–223
  corruption and, 184
  development of, 163–164
  Jan Dhan accounts, 167, 168*f*
  scams, types of, 180–182
Jan Dhan-Aadhaar-Mobile (JAM) payment pathway, 155, 167–170, 173–174, 179, 226
Janu, Swati, 46

Jeffrey, Robin, 17
Jenkins, Henry, 121–123
JFlix festival, 146
Jha, Manish, 223, 234
Jha, Rishi, 234
Jio. *See* Reliance Jio
JioCinema, 101, 102, 103*f*
Jio effect, 120
JioPlay, 100
JioSaavn, 200
Josh (video platform), 40, 119
  compared to TikTok, 121
  content creators and, 145–146, 145*f*
  influencer marketing, 151–152
  retreats and workshops for content creators, 146–148, 218
  "World Famous" initiative, 145–146, 145*f*

Kalaiselvan, Pravin, 204–205
Kaltura, 91
Kamath, Anant, 223
Kapoor, Ekta, 95, 260n27
Kar, Sohini, 197
Kar, Suman, 206–207, 209
Kaun Banega Crorepati, 181
Kauntia, Nishant, 209
Kavoori, Anandam, 17
Kaye, D. Bondy Valdovinos, 148
Keyes, James, 59, 60
Khan, Danish, 96, 114, 116
Khan, Irrfan, 1
Khan, Pehlu, 130
Khan, Samar, 94–95
"Kolaveri D.," 124
Kovacs, Anja, 228, 237
Kuaishou app, 135
KukuFM, 113
KukuTV, 113
Kumar, Neha, 51, 58, 64
Kumar, Nitish, 163
Kumar, Sangeet, 124
Kushi Gana Downloading, 46, 47*f*
Kwai, banning of, 36

Lakshmanan, Srikanth, 199–200, 206–207
Langevin, Marie, 201–202
Last-mile operators, 112
Latency, 84–85
Latour, Bruno, 57
Lenddo, 201–202
Lending platforms. *See also* Digital lending
  data as collateral, 41–42
  Pradhan Mantri MUDRA Yojana (PMMY), 26–27
  predatory microloan apps, 15–16
  software development kits (SDKs), 20–21
  women and, 13–14
Lewis, Jovan Scott, 158, 161
Limelight Networks, 103
Limits, 5, 7–8, 18
  aspirations and, 148–150
  crossing, 6
  energy and, 236
  mobile distribution and, 218–220
  regulation and, 184–185
  search for, 292n60
  thresholds as, 7
Lip-synching, 33, 123, 132, 135–136
Lithium, 235
Loan Lucky Cash, 209
Lobato, Ramon, 49
Localization, 109, 139, 151–152
Lottery scams, 158
Lotz, Amanda, 121

Mahila Money, 214
Majumdar, Shudeep, 151–152
Make in India, 27
The Making of Lynchistan: Inside India's Deadly "Gau Raksha" Network, 125

Malware, 193
Mankekar, Purnima, 178, 220
Marginalized communities. *See also* Caste system; Dalit community
 in China, 135
 community identity, 136
 Meos, 68
 TikTok and, 134–135
 trolling of, 135
Marwick, Alice, 147
McDonald, Tom, 192
Media backends, 14, 88–89, 215, 219
Media imperialism, 87
Media infrastructures, 85. See also Distribution Infrastructures.
Mediation, 123–124
Medine, David, 208
Mehta, Smith, 148
Meijas, Ulises, 230
Memes
 biological associations, 122–123
 circulation of, 122–124
 history of, 121–125
 limitlessness, sense of, 150
 political satire and, 124
 spreadability, 118–119
 on TikTok, 131–137
 vs. virals, 122
 on WhatsApp, 125–131
Memory cards, 19*f*, 45–81, 51*f*
 apps replacing, 80
 aspirational mobility, limitations of, 68–71
 aspirations and, 22
 audio players, 71, 72*f*–75*f*, 76, 77*f*
 Bhojpuri music, 45–46
 caste mobility, 71–79
 censorship, 64–66
 as circumvention technologies, 60, 62–63
 in Cuba, 52, 79–80
 cultural memory and, 64
 digital media studies of, 52
 download vendors, 18, 31, 46, 47*f*–48*f*, 53–59, 54*f*–55*f*
 informal circulation, 50–51
 as media distribution, 29, 30–31, 49–50
 mobility of digital files, 64–68
 music and, 18–19
 in North Korea, 60, 62, 81
 as part of mobile digital practices, 50–53
 personal archiving practices, 63–64
 privacy and, 62–63, 81
 vernacular languages and, 64
 videos and, 59–61
Meos, 68, 69–71, 70*f*, 256n34, 256n40
Methodologies, 9–10, 49, 89–90, 157
Mewati videos, 62, 64–66, 67*f*, 69–71, 70*f*, 78
Microcelebrities, 131, 132–133, 147–148, 153, 239–240. *See also* Content creators; Influencers
Microdramas, 113
Microentrepreneurs, 202, 221–222
Microfinance, 196–197
Microhabits, 129
Microinfluencers, 119, 121, 152–153. *See also* Influencers
Microloan apps, predatory, 41, 189–190
 alternative credit scores, 201–203, 220
 app stores and, 209
 communication networks, 193
 compared to Grameen Bank, 213–214
 data-driven algorithms, 202
 ease of deployment, 207–208
 financial companies and, 194
 in Global South, 219
 growth of, 191–192
 history of, 198–201
 inequality and, 15–16
 intermediaries, 194–195, 216

Microloan apps, predatory (cont.)
  Lenddo, 201–202
  misleading advertising on social media, 189, 209, 210*f*–211*f*
  nonbank financial companies and, 209–211
  onboarding, 189
  selfies and, 192, 207
  stakeholders and intermediaries, 207*f*
MicroSD cards. *See* Memory cards
Microsoft Azure, 89, 103
Microsoft Research India, 167
Migrant labor population, 238–239
Mills, Mara, 18
Mirzapur, 176
Mishra, Pankaj, 117–118, 152, 166, 178
Missed calls, 2
Mobile access, redistribution, 12–16
Mobile for development (M4D), 12
Mobile inclusion. *See* Digital inclusion; Financial inclusion
Mobile media, 4
  aspirational politics and, 220–221
  gender and, 237–238
  geographical mobility and, 78–79
  socioeconomic mobility and, 78–79
Mobile media distribution, 28–33. *See also* Streaming media/platforms
  affect theory and, 149
  aspirational politics and, 16, 28–33, 217
  imagined vs. actual uses of, 13
  informal operators, role of, 30–31
  intermediaries, 31
  limits of, 218–220
  memory cards and, 29, 30–31, 49–50
  to mobile phones, 10–11
  streaming and fintech in, 239
  studies of, 10, 11–12
  user-generated content, 153–154
Mobile media use, political subjectivity and, 42–43
Mobile money distribution, 167–174
Mobile money services, 11, 27, 40–41. *See also* Digital lending; Digital payments
  alternative circulatory pathways, 167–174
  demonetization and, 156–157
  fraudulent authentications, 157
  India Stack's APIs and, 30
  intermediaries and, 31–32
Mobile phones
  advertising and, 1
  affordances of, 4, 18, 57–58, 76, 181–182
  aspirational narratives, 220–221
  audience for streaming media, 91–96
  connectivity, 92–96
  cybercafes, shift from, 12
  cybercrime and, 166–167, 169, 172–173, 179
  data, costs of, 1–2
  digital divide and, 15
  emotional investment in, 76
  ergonomic aspects of, 33
  in Global South, 12
  media distribution to, 10–11
  microentrepreneurs and, 221–222
  mobility and, 19–20
  potentials vs. limitations of, 13
  range of uses and access to, 34–35
  socioeconomic mobility and, 218–219
  spread of, 17–18
  spyware and, 226
  unlimited data plans, 2–4
Mobile wala, 32, 53, 57–58, 65, 71, 85. *See also* Download vendors
Mobilities (un)limited
  aspirational/experiential level, 5
  digital infrastructures, 4–5
  politics and media distribution, 5
  socioeconomic mobility and, 218–219
  sociopolitical changes in India, 4
  techno-economic level, 5

Mobilities, aspirational, 22–23. *See also* Socioeconomic mobility
caste and, 71–79
freedom and constraints of, 81
limitations of, 68–71
mobile phones and, 4
viral vigilante videos and, 150
Modernity thesis, 16–17
Modi, Narendra, 5, 21, 184
demonetization, 155, 156
Digital India and, 36–37
social media and, 117
Mohalla Tech, 137
Mohan, Sriram, 124
Moj (video platform), 35, 40, 118, 119, 137
AI technology and, 152
audience analysis, 138–139
compared to TikTok, 121
influencer marketing, 151–152
introduction of, 138
Prajapati and, 138–145, 140*f*–141*f*, 143*f*–144*f*
retreats and workshops for content creators, 146–148, 218
M-PESA, 20, 190
mPokket, 198–199, 214
MTV, 90–91
Multichannel networks (MCNs), 31
Murthy, Gayathri, 208
Music. *See also* Vernacular music videos
on cassettes, 55–56
digitalization of, 50–51
informal economy, 46–49
memory cards and, 18–19
mobility and, 23
MusicAlly, 134
Music-streaming, 200

Nair, Adarsh, 98–100
Nath, Kailash, 201
Natile, Serena, 219
Nationalism, 27, 125, 150, 221
National Payments Corporation of India (NPCI), 37, 196
Navi Mumbai, data centers and, 232–234, 232*f*, 233*f*, 234*f*
Neoliberalism, 21, 25–26, 222–223
Neo-middle classes, 21–22, 35, 217, 220–221, 228, 247n64, 281n20, 293n67
Neomobiles, 95–96
aspirations of, 35–36
backlash against, 135
challenges of with technology, 237
commuting users, 113
entertainment vs. information for, 137
mobile digital practices, 236–240
mobile phones and, 217
neo-middle class and, 238
vernacular languages and, 40, 121, 138
Netflix, 59, 84, 100, 102, 104–105, 109, 110, 111, 112
Netflix Open Connect, 87
Net neutrality, 105
Nickelodeon, 90–91
Nilekani, Nandan, 37
Nizaruddin, Fathima, 125, 129
Nonbank financial company (NBFC), 194, 206, 207, 207*f*, 209–211
North Korea, 60, 62, 81
Notel devices, 60, 61*f*
NTT Data Center, 232, 232*f*

Oligarchy, 97
O'Neill, Jacki, 31, 80
One-time passwords (OTP), 157
Open Connect, 102, 110, 111
Open Connect Appliances (OCAs), 110
Operation WestEnd, 225
Optimization, 92, 209, 228, 252n116
Outsourcing, call centers and, 162–163
Over-the-top (OTT) aggregators, 83

Over-the-top (OTT) video services, 87, 90–91, 116. *See also* Streaming media/platforms
Oxigen, 166

Padhi, Soumendra, 176
Pahwa, Nikhil, 130
Pai, Sajith, 35
Pankaj, Ajeet Kumar, 223
Parasocial relationships, 147
Parikh, Tapan, 51
Parks, Lisa, 88–89
Patani, Disha, 132, 133*f*–134*f*
Patriarchy, 179
Paytm, 156–157, 166, 169, 187, 195
Pegasus, 226, 288n32
Pepsi, 132–133, 133*f*–134*f*
Perren, Alissa, 10
Phishing, 40–41, 155–185. *See also* Cybercrime; Jamtara, cybercrime and; *Jamtara: Sabka Number Aayega*
- call centers and, 158
- caste and gender, 174–180
- depictions of, 161, 162*f*
- digital mobile money, alternative circulatory pathways, 167–174
- introduction to, 155–160
- Jamtara's reputation and, 172–173
- regulatory limits and, 184–185
- scam calls, 160–167
- scams and labor regimes, 180–183
- youth unemployment and, 24–25

PhonePe, 156–157, 187
PicoNETS, 12, 106–109, 107*f*, 110, 115
Platform economy, 11, 23–24. *See also* Platforms
Platform imperialism, 87
Platformization, 37–38
Platform labor, 23–24. *See also* Content creators
Platform mobility, 20. *See also* Content, movement of
Platforms. *See also* Social media
- affordances of, 33, 39–40, 117, 148–149, 181–182
- audiences, search for, 217–218
- Chinese apps, 120–121
- ecosystems, 129
- intermediaries, 149, 220
- localization of, 151–152
- memes and, 118–119
- payment and media, convergence of, 239–240
- physical and logistical constraints, breaking, 219–220
- political campaigns, 117–118
- short-video platforms, 137–148
- sociality, 118–119
- spreadability, 118–119
- TikTok compared to Indian platforms, 151–152
- vernacular Indian apps, 120–121
- Western Big Tech apps, 120

Platform theory, 11–12
Plunkett, Steve, 114
Political advertising, 117–118, 152–153
Political subjectivity, mobile media use and, 15, 36, 37, 42–43, 238
Politics. *See* Aspirational politics
Poonam, Snigdha, 174, 200
Poster, Winifred, 162, 163
Pradhan Mantri MUDRA Yojana (PMMY), 26–27, 41
Pradhan Mantri Ujjwala Yojna (Prime Minister's Brightness Scheme [PMUY]), 170
Prahalad, C. K., 197
Prajapati, Raghvendra, 35, 138–142, 140*f*–141*f*, 143*f*–144*f*
Predatory inclusion, 212
Pradesh, Andhra, 123
Prime Minister's Brightness Scheme (PMUY), 170
Prime Minister's People's Money Scheme (PMJDY), 41, 155, 168, 170, 251n112

Prime Music, 200
Privacy, 93–94, 96, 225. *See also* Data collection; Surveillance
Punathambekar, Aswin, 124
Purohit, Kunal, 131
Pybus, Jennifer, 20

Raghav Kanpurwala. *See* Prajapati, Raghvendra
Rakhi, 128
Ramanathan, Arundhati, 208
Ranganathan, Nayantara, 228
Rangaswamy, Nimmi, 12, 22–23, 167
Rao, Yichen, 192
Reality TV shows, 146–147
Reliance Jio
  4G data services, 92
  advertising, 1
  arrival of, 1, 19
  aspirations and, 22
  bundling and, 83, 218
  content partnerships, 97–98
  dominance of, 97
  Hotstar, relationship with, 100–102
  streaming media and, 84
  unlimited data plans, 4, 33–34
Reserve Bank of India (RBI), 194–196, 211–212
Right-wing nationalism, 6, 27, 125, 150, 221, 267 n32, 276n33
RK, Rajeev, 107*f*, 108, 109, 110–111
Roposo app, 140
Roy, Ananya, 41
Roy, Srirupa, 225
Rural–urban divide, 40, 158–159, 179

Sachdeva, Ankush, 137
Sampson, Tony, 123
Samsung, 35
Sarkar, Bhaskar, 69
Sathe, Gopal, 199
Scams, 160–167. *See also* Cybercrime
  call centers and, 161–163
  credit card fraud, 160–161
  as networked activity, 193
  types of, 160, 180–181
  vocalizations, 163
Search engines, as intermediaries, 31
Security guards, 238–239
Selfies, and banking, 28*f*, 32–33
Self-service kiosks, memory cards and, 59–61
Sextortion, 181
Shaikh, Shadma, 136, 137, 224
Shankar, Uday, 101
ShareChat, 35–36, 137–138
SHAREit, 80
Sharma, Betwa, 128
Sharma, Supriya, 128
Sharma, Yashraj, 151
Shifman, Limor, 122
Short-video platforms, 137–148. *See also* *specific platforms*
  advertising and, 132–133
  AI technology and, 138
  audience analysis, 138–139
  content creators, recruitment of, 138–139, 142–146, 143*f*–144*f*, 145*f*, 218
  entertainment vs. information in, 137
  future of, 137–148
  history of, 124–125
  parasocial relationships and, 147–148
Shroff, Tiger, 132, 133*f*–134*f*
Side-loading, 49
Singh, Abhigyan, 45, 53, 58, 64, 69, 80
Singh, Arti, 205–206
Singh, David, 235
Singh, Moti, 128
Singhal, Vinay, 239
Skill India Digital Platform, 8–9, 27
Small media, 80

Smartphones. *See* Mobile phones
Social hierarchies, cultural limits of, 7–8, 13
Sociality, 193, 208. *See also* Spreadability; Virality
Social loan quotient (SLQ), 220
Social media. *See also* Platforms
  aspirations and, 23
  English language on, 137–138
  entertainment vs. information in, 137
  as intermediaries, 31
  marginalized communities and, 134–135
  microloan apps, misleading advertising on, 209, 210*f*–211*f*
  political campaigns, 117–118, 152–153
  vernacular languages and, 137–138, 140
Social mobility, 19, 22–23. *See also* Socioeconomic mobility
Socioeconomic mobility, 78–79, 218–219, 222–223, 238
Sociopolitical changes in India, 4
Software development kits (SDKs), 20–21, 32, 206–208, 207*f*
Sony Entertainment Television, 98
SonyLIV, 84, 87
  bundling and, 83, 99*f*, 218
  content delivery networks and, 114
  legacy broadcasters and, 98
  sports content, 100
  subscription rate, 96
  VoD services, 112
South Africa, 83, 190, 281n20, 290n43
Spitulnik, Debra, 80
Spivak, Gayatri, 69
Sports content, 100–101, 102, 103*f*
Spotify, 57
Spreadability. *See also* Virality
  factors affecting, 119–120
  introduction to, 117–121
  limits and aspirations, 148–154
  media, 121–125
  platforms vs. content, 152
  short-video platforms, 137–148
  TikTok memes, 131–137
  vigilante virality, 125–131
Spreadable Media (Jenkins), 121–122
Spyware, 226
Srinivas, S. V., 123
Star India, 102
Star Network, 90
Star Sports, 100
Star TV, 98, 100
Steinberg, Marc, 10
Stewart, Jeff, 201
Streaming media/platforms, 83–116
  actors and intermediaries, 87–88
  aesthetics for mobile phones, 113–114
  audience for, 239
  commuting users, 92–96
  conclusions, 112–116
  content delivery, 102–112
  content producers and telecoms, relationships between, 98
  distribution geographies, 95–96
  distribution infrastructure, 97
  distribution intermediaries, 88–89
  distribution networks, 86–90
  distribution networks, tracing, 88–89
  encoding process, 2–3, 86, 92–93
  Hotstar and Jio, collaboration and competition, 100–102
  introduction to, 83–87
  mobile phone audience, 91–96
  platforms, growth of, 84
  privacy and, 93–94, 96

shift to, 112
sports content, 100–102, 103*f*
starting platform for, 90–91
telecom bundling, 97–100, 99*f*, 218
unbundling, 100–102
vernacular languages and, 239
Subaltern-popular representations, 68–71
Subin, 66–67, 66*f*
Subin sher Chandanki ke (Subin, the lion of Chan danki), 66–67, 66*f*
Subscription video-on-demand (SVoD), 86, 87, 101–103, 109, 239
Sundaram, Ravi, 15, 53, 125
Super apps, 98, 261n34
Surveillance, 33, 193, 199, 200. *See also* Data collection
access and, 224–230
anticorruption and, 225
dataveillance, 229
demonetization and, 226
Global South and, 225
by mobile phone users, 225
public attitudes toward, 213, 224–225
social, 237–238
technologies for, 226
tracking software, 200–201
user sociality and, 208
Swartz, Lana, 16, 20

Tacchi, Jo, 13
Talent agents, 31, 142–146, 143*f*–144*f*
Tapzo, 166, 169
Tarde, Gabriel, 123, 128
Taxi drivers, 63–64, 65, 69
Technocapitalism, 5–6
Techno-nationalism, 21
Techno-populism, 150
Tehelka magazine, 124–125
Telecom bundling. *See* Bundling
Telecommunication companies. *See also specific companies*
audiences, search for, 217–218
bundling and, 97–100, 218
mobile phone audience, 91–96
in streaming distribution networks, 87–90
SVoD companies, relationships with, 101–102
unbundling, 100–102
Telecom revolution in India, 21
Telegram app, 181
Tenhunen, Sirpa, 237
Text messaging services (SMS), 125
Thomas, Pradip, 36
Three Wheels United, 239
Thresholds, 7–8
Throttling, 3
TikTok, 40
advertising, 132–134, 133*f*–134*f*, 150
aspirations and, 22
banning of, 36, 121, 138, 152
Bollywood films and, 131–132
challenges, 132, 134
China and, 135, 136
compared to Indian platforms, 151–152
dance videos, 132–134, 133*f*–134*f*
features that lead to virality, 120
hate, spreading on, 135–136, 150
marginalized communities and, 134–135
memes and, 131–137
mobile media distribution and, 33
political campaigns, 118
popularity of, 121
positive aspects of, 131
trolling of marginalized communities, 135
vigilante groups, 131
Tinic, Serra, 121

TMW, 166
Tongdun, 208
Toyama, Kentaro, 167
Tracking software, 200–201
Tripathy, Ratnakar, 78
Trolling, 135
Truck drivers, 65–66, 67–68, 67*f*
TryCash, 191
Tryon, Chuck, 3, 20
Tussey, Ethan, 93–94

Uber, 56
Udupa, Sahana, 125
Unemployment, 164. *See also* Youth unemployment
Unified Payments Interface (UPI), 155–156, 167–168, 191, 196, 214–215, 239
Unique identification system (UID), 155, 167. *See also* Aadhaar database
Unlimited, 7, 16, 242n10. *See also* Limits
  affect and, 6
  data plans, 2–4, 33–34, 34*f*
  as descriptor, 2
  as heuristic, 2
  Netflix and, 59, 255n25
  as promise, 5
  telecom ads and, 1, 34
Upadhyay, Vikas, 153
User-generated content, 118–119, 148–154

Van der Veer, Peter, 27–28
Varma, Pramod, 227
Vaz, Rahul, 90, 91–93
Venkatanarayanan, Anand, 217–218
Vernacular identity, 68
Vernacular languages, 45, 64, 124, 137–138, 140, 153–154, 239
Vernacular music videos, 30–31, 39, 49, 52–53, 58, 131
Vertical video platforms, 113
Viacom18, 90–91, 101, 102
Video-editing features, 123
Video-on-demand (VoD) streaming services, 83, 112. *See also* Streaming media/platforms
Video platforms, 38. *See also specific platforms*
Vigilante groups, 6, 125–131, 149–150
Vijeeta, 147–148
Virality. *See also* Spreadability
  advertising and, 132–134, 150
  affordances and, 120, 130
  app features and, 120
  of messages, information, and rumor, 120, 149–150
  moods, feelings, and emotions, 123
  platforms vs. content, 152
  user/audience effect on, 121–122
Viral media thesis, 121–122
Virals, vs. memes, 122
Virtual gifts, 239–240
Vishing. *See* Cybercrime
Vodacom, 83
Vodafone, 1, 97–98
Voot, 90–91, 102

Wallis, Cara, 17
Weibo (microblog platform), 135
WhatsApp
  affect, 129–130
  aspirations and, 22
  end-to-end encryption, 125
  features that lead to virality, 120
  forwarding practices, 129–130
  lending platforms and, 15–16
  mobile media distribution and, 33
  political campaigns, 117–118
  predatory microloan apps and, 192
  primary sharing app, 80
  vernacular music videos and, 131

vigilante groups, 6, 125–131, 126*f*, 127*f*, 149–150
virality of messages, information, and rumor, 120, 149–150
Witteborn, Saskia, 70
Winner, Langdon, 292n60
Women. *See* Gender
World Bank, 184, 227

Xiaomi, 35

Yadav, Anumeha, 169
Yadav, Khesari Lal, 35, 139
Yotta Infrastructure, 232, 233*f*
Youth unemployment, 8–9, 24–25, 27, 170. *See also* Unemployment
YouTube, 56, 118, 130–131, 153, 189, 209
Yunus, Muhammad, 196–197

ZEE5, 84
Zip Dial, 2
Zulli, David James, 132
Zulli, Diana, 132

Publisher contact:
The MIT Press
Massachusetts Institute of Technology
77 Massachusetts Avenue, Cambridge, MA 02139
mitpress.mit.edu

EU Authorised Representative:
Easy Access System Europe, Mustamäe tee 50,
10621 Tallinn, Estonia
gpsr.requests@easproject.com

Printed by Integrated Books International,
United States of America